An Introduction to the History of Algebra

Solving Equations from Mesopotamian Times to the Renaissance

MATHEMATICAL WORLD VOLUME 27

An Introduction to the History of Algebra

Solving Equations from Mesopotamian Times to the Renaissance

Jacques Sesiano

Translated by
Anna Pierrehumbert

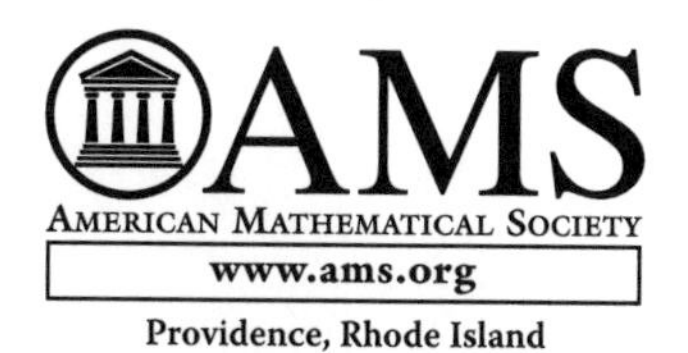

Translated by Anna Pierrehumbert

2000 *Mathematics Subject Classification*. Primary 01A05, 17–XX, 20–XX, 30–XX, 35–XX, 40–XX.

For additional information and updates on this book, visit
www.ams.org/bookpages/mawrld-27

Library of Congress Cataloging-in-Publication Data

Sesiano, Jacques.
[Introduction à l'histoire de l'algèbre. English]
An introduction to the history of algebra : solving equations from Mesopotamian times to the Renaissance / Jacques Sesiano ; translated by Anna Pierrehumbert.
p. cm. – (Mathematical world ; v. 27)
Includes index.
ISBN 978-0-8218-4473-1 (alk. paper)
1. Algebra—History. I. Title.

QA151.S4513 2009
512.009–dc22 2009008068

∞ The paper used in this book is acid-free and falls within the guidelines
established to ensure permanence and durability.
Visit the AMS home page at `http://www.ams.org/`

10 9 8 7 6 5 4 3 2 1 14 13 12 11 10 09

Contents

Preface vii

Chapter 1. Algebra in Mesopotamia 1
1.1. Introduction 1
1.2. Linear Systems 4
1.3. Quadratic Equations and Systems 9

Chapter 2. Algebra in Ancient Greece 17
2.1. Introduction 17
2.2. Common Algebra 19
2.3. Diophantine Algebra 31

Chapter 3. Algebra in the Islamic World 53
3.1. Introduction 53
3.2. Al-Khwārizmī 56
3.3. Abū Kāmil 63
3.4. The Geometric Construction of Solutions of the Quadratic Equation 79
3.5. The Cubic Equation 80

Chapter 4. Algebra in Medieval Europe 93
4.1. Introduction 93
4.2. The *Liber mahameleth* 95
4.3. Leonardo Fibonacci 101
4.4. Later Developments 115

Chapter 5. Algebra in the Renaissance 123
5.1. The Development of Algebraic Symbolism 123
5.2. The General Solution of the Cubic and Quartic Equations 125
5.3. The Solution of the Cubic Equation in Italy 129
5.4. The Solution of the Quartic Equation in Italy 134
5.5. Bombelli and Imaginary Numbers 135
5.6. *Casus irreducibiles* 140

Appendix A. Mesopotamian Texts in Translation 143

Appendix B. Greek and Latin Texts 147

Appendix C. Arabic Texts 157

Appendix D. Hebrew Text 163

Appendix E. French, German, Italian, and Provençal Texts 165

Index 171

Preface

The years following 2000 B.C. and the years preceding A.D. 1600 mark two important turning points in the development of algebra. The first witness the solution of equations and systems of the first and second degrees; to the second is associated the solution of the third- and fourth-degree equations, which are also the last in which the unknown may be expressed by means of a general formula. The years between these two turning points also saw the successive extensions of the number system: confined initially to the positive integers and fractions, it next came to include positive irrational numbers; then, towards the end of the Middle Ages, negative numbers made a timid appearance, and this was followed a little later by the first use of complex numbers.

This book does not aim to give an exhaustive survey of the history of algebra up to early modern times, but merely to present some significant steps in solving equations and, wherever applicable, to link these developments to the extension of the number system. Various examples of problems, with their typical solution methods, are analyzed, and sometimes translated completely. Indeed, it is another aim of this book to ease the reader's access to modern editions of old mathematical texts, or even to the original texts; to this end, some of the problems discussed in the text have been reproduced in the Appendix in their original language (Greek, Latin, Arabic, Hebrew, French, German, Provençal, and Italian) with explicative notes.

Bibliographical references have been restricted to the particular subjects considered. General information may be found in classical textbooks on the history of mathematics, such as K. Vogel's *Vorgriechische Mathematik II*, Hannover/Paderborn 1959 (Mesopotamian mathematics), T. Heath's two-volume *History of Greek mathematics*, Oxford 1921 (reprinted), M. Cantor's four-volume *Vorlesungen über Geschichte der Mathematik*, Leipzig 1900–13 (reprinted), J. Tropfke's seven-volume *Geschichte der Elementar-mathematik in systematischer Darstellung*, Leipzig 1921–24 and (vol. 1-4 only) 1930–40.

While preparing this English edition of the original French book published by the Presses polytechniques et universitaires romandes, I had the pleasure to work with Dr. Ina Mette, in Book Acquisitions at the American

Mathematical Society, and with Anna Pierrehumbert, in charge of the translation and typesetting of the book. I wish to express to them my sincerest thanks for their patience and kindness. I also wish to express my indebtedness to my student Christophe Hebeisen, who has been of constant help whenever technical difficulties appeared in relation to the production of the text.

Lausanne, Switzerland
September, 2008

Chapter 1

Algebra in Mesopotamia

1.1. Introduction

Most of the mathematical cuneiform texts we have today date from about 1800 B.C. They are written on clay tablets in Akkadian, the language of the Semitic people who conquered the land of the Sumerians, in the modern Iraq, around 2000 B.C. These invaders adopted both the culture and the writing of the Sumerians. They also retained the knowledge of the Sumerian language, which would most likely have remained unintelligible to us without the Sumerian–Akkadian lexicons that they compiled and that have been preserved to the present day.[1]

It is to the Akkadians, too, that we owe the transmission of that outstanding feature of Sumerian culture, the development of mathematics. But this was not known until modern times: first came the deciphering of cuneiform writing, in the nineteenth century, then the study of cuneiform mathematical texts, in the first half of the twentieth century. Although these texts were written in Akkadian, they continued to use Sumerian words for technical terms. What was thus presumed to be a Sumerian heritage then revealed the existence of a quite particular form of algebra with pairs of linear equations as well as quadratic equations and systems being solved, with surprising ease, by applying a few elementary identities. But before going any further, we must first consider another particularity of these texts, which has to do with the number system in use at the time.

The earliest Sumerian clay tablets date from around 3200 B.C. From these we see that, even before the development of an alphabet, a primitive number system involving both a decimal base and a sexagesimal base was already in use. Thus D, ○, D, ◯ represented, respectively, 1, 10, 60, 3600; then, with the sign for 10 combined with D and ◯, signs for 600 and 36000

[1]The grammatical structure of Sumerian, which is an agglutinative language, appears to place it in the family of Ural-Altaic languages, along with Finnish, Hungarian, and Turkish. But this would have been of no help in understanding the language itself. With Akkadian, which is closely related to the other Semitic languages, such a problem did not arise.

were obtained. The other numbers were formed by repetition of these six symbols. That base 10 was used is not surprising: throughout the world, the ten human fingers had long provided a primitive method of counting. However, no such obvious reason can account for the choice of base 60. Modern speculations about the numerous divisors of the integer 60 appear to overlook the age of the earliest evidence for this system. On the tablets mentioned above, the symbol used for 60 appears alongside primitive drawings from before the invention of writing—that is, from a time when people could hardly have been considering the divisibility of numbers. A more reasonable explanation might be found in connection with the fundamental requirements of astronomy or the calendar. Perhaps, for example, the use of 60 had to do with the approximate number of the days and nights in a lunar month.

Whatever its origin may be, the above sexagesimal system, and thus the sexagesimal base, continued to be used after the development of written language. Around the middle of the third millenium B.C., a similar but improved numeral system came into being. Two symbols, one for unity (𒁹) and one for ten (𒌋), were used to compose the sexagesimal digits from 1 to 59 via repetition. Thus 𒐈, 𒌋𒁹, 𒌍𒐈 represent 3, 11, 33 (with the three unit symbols stacked), respectively. Note that, unlike the digits we use, these sexagesimal digits were of widely varying length. For purely practical reasons, the system evolved into a "place-value system" for expressing numbers above 59, that is, one in which the value of each symbol depends on its place within the number. The sexagesimal digits were simply juxtaposed and, when read from left to right, indicated the coefficients of decreasing (positive or negative) powers of 60. Thus 𒐈 𒌋𒁹 𒌍𒐈 represents the three-digit number $3'11'33$.

So far, this is just like our system except that we have 9 significant (i.e. nonzero) digits, whereas they had 59. Now such a place-value system must have some sign to indicate an empty space, that is, in the present case, the absence of a certain power of 60; but at the time, and for more than two millenia (till about 300 B.C.), there was no proper symbol for zero. An intermediate zero was marked simply by an empty space on the tablet, a gap in the writing. Now this could easily lead to confusion: first, no precise length can be attributed to such a gap since the lengths of the sexagesimal digits are not uniform—and thus two consecutive zeros could appear to be a single one; second, an empty space could be left merely to accommodate a rough or uneven surface of the clay tablet. With such an inherent weakness, this system can hardly have been created deliberately. But there is another, even more troublesome, weakness. Since an empty space is to be inserted between two sexagesimal digits, it is impossible to represent a number with initial or final zeros. Consequently, all numbers in cuneiform script, whether consisting of one digit or more, are defined only up to a factor equal to some (positive or negative) power of 60—which is as if our 132 were supposed to be preceded or followed by an indeterminate number of zeros. This should have been an obstacle to any development of mathematics. In actual fact, in mathematical problems this inconvenience was often avoided by mentioning at the outset what units of measurement were being used. This helped to determine the order of magnitude of the given values, whereby the second weakness of the system was overcome.

Learning those 59 digits was no more difficult than learning our nine is; with only two symbols to be repeated as necessary, it was in fact less so. It was thus not in writing numbers but in calculating that the use of a sexagesimal system led to difficulty. Performing calculations with our base 10 requires knowing 36 products of one digit by another, not counting products where one of the factors is 0 or 1 or those made superfluous by commutativity. On the other hand, after eliminating banal and repetitive products in the same way, base 60 requires knowing 1711 products. As it would hardly have been reasonable to expect people to learn them all by heart, written multiplication tables were needed for performing calculations. The abundance of such tables in museums today attests both to the disadvantages of the sexagesimal system and to its widespread use in calculating.

Still, even with the help of tables, much mental agility was required of Mesopotamian mathematicians. The results of the multiplication of two numbers with several sexagesimal digits were given directly in the solutions of problems, which means that after extracting all the necessary products of two sexagesimal digits from the tables, Mesopotamians were able to mentally add the digits occupying the same sexagesimal places or, at least, to verify the result. Thus, in one of the examples we shall see below (page 12), $14'30$ by $14'30$ is directly given as $3'30'15$.[2] On the other hand, division was reduced to multiplication by the inverse of the divisor; this required first consultation of tables indicating, for a given integer less than 60, its inverse and, second, multiplication tables indicating the products by the inverse in question[3]. Finally, note that there were also tables giving exact roots of squares (as well as some approximate roots of non-square integers). This is why, in one of our examples (page 10), it is just said that $14'30'15$ is the square of $29'30$.[4] In short, mathematics in Mesopotamia was clearly inextricably linked to the existence of auxiliary tables.

As inconvenient as it was, the sexagesimal system was used in Mesopotamia for two millenia, and astronomers in particular continued to use it to record their observations. Ptolemy (A.D. 150) mentions in his fundamental text on astronomy, which was to be influential throughout the Middle Ages, that since the beginning of the reign of Nabonassar (747 B.C.), "the ancient (Mesopotamian) observations are, on the whole, preserved down to

[2] In our writing: $14 + \frac{30}{60}$ by $14 + \frac{30}{60}$ is $3 \cdot 60 + 30 + \frac{15}{60}$, that is, $(14 + \frac{1}{2})^2 = 210 + \frac{1}{4}$.

[3] Assuming that the inverse had a finite sexagesimal expansion, rather than an infinite periodic one. In base 60, this is the case for all numbers of the form $2^\alpha 3^\beta 5^\gamma$, just as the inverses of numbers of the form $2^\alpha 5^\beta$ have a finite expansion in base 10. In general, the necessary and sufficient condition for a number to have an inverse with finite expansion in a given base is that the representation of the number as a product of primes contains only prime factors of that base.

[4] In our writing: $14 \cdot 60 + 30 + \frac{15}{60}$ $(= 870 + \frac{1}{4})$ is the square of $29 + \frac{30}{60}$ $(= 29 + \frac{1}{2})$.

our own time."[5] Several centuries of continuous observations provided an invaluable body of data for the computation of planetary periods, and Greek scientists knew how to best make use of this information. As it would have been an overwhelming task to convert all these data into the decimal system they used, the Greeks maintained the sexagesimal system for astronomical measurements, including measurements of time and angles. This was also adopted by the Indians as early as antiquity, probably via the Greeks. Then, together with Indian and Greek astronomy, it reached the Muslims, who in turn transmitted this notation to medieval Christian Europe. The sexagesimal division of degrees and hours still used today is thus a living witness to the sexagesimal base once used by the Sumerians already in prehistoric times.

1.2. Linear Systems

Consider a linear system of one of the two forms

$$\begin{cases} ax + by = k \\ x \pm y = l, \end{cases}$$

where a, b, k, and l are given quantities with k and l positive and at least one of a and b positive.

To solve this system, one could either

(1) express one of the unknown quantities in terms of the other by using the second equation, introduce this into the first equation, and solve the new equation $ax \pm b(l - x) = k$,

or

(2) take as new unknown $z = \frac{x-y}{2}$ (or $\frac{x+y}{2}$), and substitute $\frac{x+y}{2} \pm \frac{x-y}{2} = \frac{l}{2} \pm z$ (respectively $z \pm \frac{l}{2}$) for x and y in the first equation.

Using the second method to solve this system may well seem odd to modern readers. However, the Mesopotamians resorted to the second as much as to the first, as the following two examples will illustrate.

These two examples follow one another on the same tablet, and their close relationship is visible both in their form and in their similar givens and solutions. We will first present the original text in translation, then a summary of the computations involved, and finally explain the solution. Note that the sexagesimal numbers have been converted to decimal numbers in the translations; see Appendix A for the same texts with sexagesimal

[5] See Ptolemy's *Almagest*, III, 7. Of the Greek text published by J. Heiberg, *Syntaxis mathematica* (2 vol.), Leipzig 1898–1903, we have a German translation by K. Manitius (*Des Claudius Ptolemäus Handbuch der Astronomie* (2 vol.), Leipzig 1912–13; reprint: Leipzig 1963) and an English translation with commentary by G. Toomer (*Ptolemy's Almagest*, New York 1984). A French translation with the Greek text, in a less critical edition, exists by the Abbé N. Halma, *Composition mathématique de Claude Ptolémée* (2 vol.), Paris 1813–16 (reprint: Paris 1988).

numbers and, in bold face, the initial and final zeros that, as mentioned above, were not indicated in Mesopotamian texts[6].

In these two problems, we are asked to find the areas of two fields of grain knowing their respective yields per unit area, the difference in the total yields, as well as the sum of their areas (first problem) or the difference of their areas (second problem). The units employed are the *bur*, which equals 1800 *sar* (1 *sar* $\cong$ 36 m^2), for surface area; and the *kur*, which equals 300 *sila* (1 *sila* $\cong$ 1 liter), for volume. In both problems the respective yields per unit area of the two fields are the same: 4 *kur* = 1200 *sila* per *bur* for the first field and 3 *kur* = 900 *sila* per *bur* for the second. The difference in their total yields, 500 *sila*, is the same as well. The sum of the areas is given to be 1800 *sar* in the first problem, while the difference is given to be 600 *sar* in the second. Finally, note that in each text the statement of the problem is followed by a change of units (a welcome step in the absence of final zeros; see page 2) and that the answer is verified after it has been calculated.

Example 1. Tablet 8389, Museum of the Ancient Near East, Berlin, Problem 1 [Appendix A.1][7]

(α) *Per bur, I obtained* 4 *kur of grain. Per second bur, I obtained* 3 *kur of grain. One grain exceeds the other by* 500. *I added my fields;* (*it gives*) 1800. *What are my fields?*

Put 1800, *the bur. Put* 1200, *the grain he obtained. Put* 1800, *the second bur. Put* 900, *the grain he obtained.*[8] *Put* 500, *that by which one grain exceeds the other. And put* 1800, *the sum of the areas of the fields.*

Next, divide in two 1800, *the sum of the areas of the fields*[9]*;* (*it gives*) 900. *Put* 900 *and* 900, *twice. Take the inverse of* 1800, *the bur;* (*it gives*) $\frac{1}{1800}$.[10] *Multiply* $\frac{1}{1800}$ *by* 1200, *the grain he obtained;* (*it gives*) $\frac{2}{3}$, *the false grain. Multiply* (*it*) *by* 900, *which you have put twice*[11]*;* (*it gives*) 600. *Let your head hold* (*it*)[12]. *Take the inverse of* 1800, *the second bur;* (*it gives*) $\frac{1}{1800}$. *Multiply* $\frac{1}{1800}$ *by* 900, *the grain he obtained*[13]*;* (*it gives*) $\frac{1}{2}$, *the false*

---

[6]The translation of these texts follows that of F. Thureau-Dangin, *Textes mathématiques babyloniens*, Leiden 1938. For a (more literal) English translation of these texts, see J. Høyrup, *Lengths, Widths, Surfaces: a Portrait of Old Babylonian Algebra and Its Kin*, New York 2002. The particular algebraic way of solving equations in Mesopotamian mathematics was first noted by Kurt Vogel, "Zur Berechnung der quadratischen Gleichungen bei den Babyloniern", *Unterrichtsblätter für Mathematik und Naturwissenschaften*, 39 (1933), pp. 76–81 (reprinted pp. 265–273 in J. Christianidis (ed.), *Classics in the History of Greek Mathematics*, Dordrecht 2004).

[7]In all translations of texts, the words in parentheses have been added by us for clarity.

[8]Conversion of *bur* to *sar* and *kur* to *sila*.

[9]This specification is welcome, as there is also another 1800 (the *bur* converted into *sar*).

[10]Note that this way of writing fractions is anachronistic, as all fractions were expressed sexagesimally.

[11]and not the 900 that is the yield of the second field.

[12]This means that an intermediate result has been obtained that will be used later. A similar phrase is found in medieval mathematics (see note 65).

[13]and not the 900 of the half-area.

grain. Multiply (*it*) *by* 900, *which you have put twice;* (*it gives*) 450. *By what does* 600, *which your head holds, exceed* 450*? It exceeds* (*it*) *by* 150. *Subtract* 150, *that by which it exceeds, from* 500, *that by which one grain exceeds the other; you leave* 350. *Let your head hold* 350, *which you left. Add the coefficient* $\frac{2}{3}$ *and the coefficient* $\frac{1}{2}$; (*it gives*) $\frac{7}{6}$. *I do not know its inverse*[14]. *What must I put to* $\frac{7}{6}$ *to have* 350, *that your head holds? Put* 300. *Multiply* 300 *by* $\frac{7}{6}$; *this gives you* 350. *From* (*one*) 900, *which you have put twice, subtract, and to the other add,* 300, *which you have put; the first is* 1200, *the second* 600. *The area of the first field is* 1200, *the area of the second field* 600.

If the area of the first field is 1200 *and the area of the second field is* 600, *what is their grain? Take the inverse of* 1800, *the bur;* (*it gives*) $\frac{1}{1800}$. *Multiply* $\frac{1}{1800}$ *by* 1200, *the grain he obtained*[15]; (*it gives*) $\frac{2}{3}$. *Multiply* (*it*) *by* 1200, *the area of the first field;* (*it gives*) 800, (*which is*) *the grain of* 1200, *the area of the first field. Take the inverse of* 1800, *the second bur;* (*it gives*) $\frac{1}{1800}$. *Multiply* $\frac{1}{1800}$ *by* 900, *the grain he obtained;* (*it gives*) $\frac{1}{2}$. *Multiply* $\frac{1}{2}$ *by* 600, *the area of the second field;* (*it gives*) 300, (*which is*) *the grain of* 600, *the area of the second field. By what does* 800, *the grain of the first field, exceed* 300, *the grain of the second field? It exceeds* (*it*) *by* 500.

(β) The numerical calculations used to solve the problem and to verify the answer found are successively (with the conversions immediately following the statement omitted here):

$$1800/2 = 900$$

$$900$$

$$900$$

$$\begin{cases} 1800^{-1} = \frac{1}{1800} \\ 1200 \cdot \frac{1}{1800} = \frac{2}{3} & \text{“false grain”} \\ 900 \cdot \frac{2}{3} = 600 & \text{“hold”} \end{cases}$$

$$\begin{cases} 1800^{-1} = \frac{1}{1800} \\ 900 \cdot \frac{1}{1800} = \frac{1}{2} & \text{“false grain”} \\ 900 \cdot \frac{1}{2} = 450 \end{cases}$$

$$600 - 450 = 150$$

$$500 - 150 = 350 \quad \text{“hold”}$$

$$\frac{2}{3} + \frac{1}{2} = \frac{7}{6}$$

[14]The inverse of 7 is not found in the usual tables for inverses (nor in the multiplication tables), as it does not have a finite sexagesimal expansion (see note 3).

[15]and not the area of the first field.

$\frac{6}{7}$ "not known"

Now $\frac{7}{6} \cdot z = 350$ for $z = 300$

(The text verifies this.)

$900 + 300 = 1200$ "area of the first field"

$900 - 300 = 600$ "area of the second field"

$$\begin{cases} 1200 \cdot \frac{1200}{1800} = 800 & \text{"grain of the first field"} \\ 600 \cdot \frac{900}{1800} = 300 & \text{"grain of the second field"} \\ \text{the difference is, indeed, 500.} \end{cases}$$

(γ) Interpretation:

First, $\frac{1200}{1800} = \frac{2}{3}$ and $\frac{900}{1800} = \frac{1}{2}$ are the yields (in *sila*, not in *kur*) of the first and second fields per unit area (*sar*, not *bur*), respectively, which the tablet refers to as the "false grain." The "true grain" would then be the total production of each field, thus $\frac{2}{3}x$ and $\frac{1}{2}y$ if x and y represent the areas of the two fields. Therefore, in modern mathematical notation, we are asked to solve the system

$$\begin{cases} x + y = 1800 \\ \frac{2}{3}x - \frac{1}{2}y = 500. \end{cases}$$

We already know that $\frac{x+y}{2} = 900$. We now take, to replace x and y, the auxiliary unknown $z = \frac{x-y}{2}$. By adding and subtracting these, we obtain $x = 900 + z$ and $y = 900 - z$, and the second equation becomes

$$\frac{2}{3}(900 + z) - \frac{1}{2}(900 - z) = 500.$$

Looking back at the computations given in the text, we see that they correspond to calculating

$$\frac{2}{3}900 - \frac{1}{2}900 + \frac{2}{3}z + \frac{1}{2}z = 500,$$

$$600 - 450 + \frac{7}{6}z = 500,$$

$$150 + \frac{7}{6}z = 500,$$

$$\frac{7}{6}z = 500 - 150 = 350.$$

With $z = 300$, we then find, as in the text, that $x = 900 + z = 1200$ and $y = 900 - z = 600$.

In summary, what the text does is replace the two unknowns by a single unknown and then solve the resulting equation. But all this is done without the slightest explanation. The only comments given in the text are to specify the concrete meaning of the quantities involved, particularly in order to differentiate between two equal numerical values attributed to different quantities, and to point out some problematic step in the computations (see the footnotes).

Example 2. Tablet 8389, Museum of the Ancient Near East, Berlin, Problem 2 [Appendix A.2]

Per bur, I obtained 4 *kur of grain. Per second bur, I obtained* 3 *kur of grain. Now (there are) two fields. One field exceeds the other by* 600. *One grain exceeds the other by* 500. *What are my fields?*

Put 1800, *the bur. Put* 1200, *the grain he obtained. Put* 1800, *the second bur. Put* 900, *the grain he obtained. Put* 600, *that by which one field exceeds the other. And put* 500, *that by which one grain exceeds the other.*

Take the inverse of 1800, *the bur; (it gives)* $\frac{1}{1800}$. *Multiply (it) by* 1200, *the grain he obtained; (it gives)* $\frac{2}{3}$, *the false grain. Multiply* $\frac{2}{3}$, *the false grain, by* 600, *that by which one field exceeds the other; (it gives)* 400. *Subtract (it) from* 500, *that by which one grain exceeds the other; you leave* 100. *Let your head hold* 100, *that you left. Take the inverse of* 1800, *the second bur; (it gives)* $\frac{1}{1800}$. *Multiply* $\frac{1}{1800}$ *by* 900, *the grain he obtained; (it gives)* $\frac{1}{2}$, *the false grain. By what does* $\frac{2}{3}$, *the false grain, exceed* $\frac{1}{2}$, *the false grain? It exceeds (it) by* $\frac{1}{6}$. *Take the inverse of* $\frac{1}{6}$, *that by which it exceeds; (it gives)* 6. *Multiply* 6 *by* 100, *which your head holds; (it gives)* 600, *the area of the first field. To* 600, *the area of the field, add* 600, *that by which one field exceeds the other; (it gives)* 1200. *The area of the second field is* 1200.

If the area of the first field[16] *is* 1200 *and the area of the second field (is)* 600, *what is their grain? Take the inverse of* 1800, *the bur; (it gives)* $\frac{1}{1800}$. *Multiply* $\frac{1}{1800}$ *by* 1200, *the grain he obtained; (it gives)* $\frac{2}{3}$. *Multiply (it) by* 1200, *the area of the (first) field; (it gives)* 800, *(which is) the grain of* 1200, *the area of the (first) field. Take the inverse of* 1800, *the second bur; (it gives)* $\frac{1}{1800}$. *Multiply* $\frac{1}{1800}$ *by* 900, *the grain he obtained; (it gives)* $\frac{1}{2}$. *Multiply (it) by* 600, *the area of the second field; (it gives)* 300, *(which is) the grain of* 600, *the area of the (second) field. By what does* 1200, *the area of the (first) field, exceed* 600, *the area of the second field? It exceeds (it) by* 600. *By what does* 800, *the (first) grain, exceed* 300, *the second grain? It exceeds (it) by* 500.

With the same notation as before, the system of equations involved in the second problem is

$$\begin{cases} x - y = 600 \\ \frac{2}{3}x - \frac{1}{2}y = 500. \end{cases}$$

We could set $z = \frac{x+y}{2}$ as new unknown and solve the problem as before. The text, however, proceeds as we would, by substituting $x = y + 600$ from the first equation into the second. This yields

$$\frac{2}{3}(y + 600) - \frac{1}{2}y = 500.$$

[16]The "second field" above, the "first field" in the verification of the previous problem.

The subsequent steps in the text correspond to simplifying this equation and solving for y; indeed, the author computes

$$\frac{2}{3} \cdot 600 = 400,$$

then

$$500 - 400 = 100,$$

and finally

$$y = \frac{100}{\frac{2}{3} - \frac{1}{2}} = 6 \cdot 100 = 600.$$

1.3. Quadratic Equations and Systems

From Mesopotamian times to the Renaissance, quadratic equations usually appeared in one of three standard forms, differing according to the place of each of the three terms, all positive, on either side of the equality. The equation $ax^2+bx+c=0$ with $a, b, c > 0$ did not appear, and could not appear, as only equations with positive (and rational) solutions were considered. Thus the three possible cases are

(1) $ax^2 + bx = c$, with solution

$$x = \frac{-\frac{b}{2} + \sqrt{\left(\frac{b}{2}\right)^2 + ac}}{a};$$

or, equivalently, in reduced form

$$x^2 + px = q,$$

with solution

$$x = -\frac{p}{2} + \sqrt{\left(\frac{p}{2}\right)^2 + q}\,.$$

(2) $ax^2 = bx + c$, with solution

$$x = \frac{\frac{b}{2} + \sqrt{\left(\frac{b}{2}\right)^2 + ac}}{a};$$

or (in reduced form)

$$x^2 = px + q,$$

with solution

$$x = \frac{p}{2} + \sqrt{\left(\frac{p}{2}\right)^2 + q}\,.$$

(3) $ax^2 + c = bx$, with the two possible solutions

$$x = \frac{\frac{b}{2} \pm \sqrt{\left(\frac{b}{2}\right)^2 - ac}}{a};$$

or (in reduced form)

$$x^2 + q = px,$$

with solutions

$$x = \frac{p}{2} \pm \sqrt{\left(\frac{p}{2}\right)^2 - q}\,.$$

Only the last of these cases has two positive solutions—as long as the discriminant is positive.

Some of the preserved Mesopotamian tablets clearly demonstrate a knowledge of the first two formulas. This is illustrated by the next two examples, where, as before, we present the text followed by its mathematical interpretation.

Example 1. Tablet 13901, British Museum, Problem 1 [Appendix A.3]

I added the area and the side of my square; (it gives) $\frac{3}{4}$.

You put 1, *the unit. You divide in two* 1; *(it gives)* $\frac{1}{2}$. *You multiply* ($\frac{1}{2}$) *by* $\frac{1}{2}$; *(it gives)* $\frac{1}{4}$. *You add* $\frac{1}{4}$ *to* $\frac{3}{4}$; *(it gives)* 1. *It is the square of* 1. *You subtract* $\frac{1}{2}$, *which you multiplied, from* 1; *(it gives)* $\frac{1}{2}$, *the side of the square.*

The equation here is $x^2 + x = \frac{3}{4}$, which is of the form $x^2 + px = q$. The calculations performed (along with their symbolic equivalents) are

$1\ (=p)$	p
$\downarrow /2$	
$\frac{1}{2}$	$\frac{p}{2}$
$\downarrow$ square	
$\frac{1}{4}$	$\left(\frac{p}{2}\right)^2$
$\downarrow +\frac{3}{4}$	
1	$\left(\frac{p}{2}\right)^2 + q$
$\downarrow$ square root	
1	$\sqrt{\left(\frac{p}{2}\right)^2 + q}$
$\downarrow -\frac{1}{2}$	
$\frac{1}{2}$	$\sqrt{\left(\frac{p}{2}\right)^2 + q} - \frac{p}{2} = x.$

The reader may not be fully convinced by this solution, as the quantity 1 we started with is both the coefficient of x^2 and that of x. But later we will see an example of the solution of an equation of this type where the coefficient of x^2 is not 1.

Example 2. Same tablet, Problem 2 [Appendix A.4]

I subtracted from the area the side of my square; (it gives) 870.

You put 1, *the unit. You divide in two* 1; *(it gives)* $\frac{1}{2}$. *You multiply* ($\frac{1}{2}$) *by* $\frac{1}{2}$; *(it gives)* $\frac{1}{4}$. *You add (it) to* 870; *(it gives)* $870 + \frac{1}{4}$. *It is the square of* $29 + \frac{1}{2}$. *You add* $\frac{1}{2}$, *which you multiplied, to* $29 + \frac{1}{2}$; *(it gives)* 30, *the side of the square.*

The equation, $x^2 - x = 870$, is of the form $x^2 = px + q$. The text performs the following calculations:

$1\ (= p)$	p
$\downarrow /2$	
$\frac{1}{2}$	$\frac{p}{2}$
$\downarrow$ square	
$\frac{1}{4}$	$\left(\frac{p}{2}\right)^2$
$\downarrow +870$	
$870 + \frac{1}{4}$	$\left(\frac{p}{2}\right)^2 + q$
$\downarrow$ square root	
$29 + \frac{1}{2}$	$\sqrt{\left(\frac{p}{2}\right)^2 + q}$
$\downarrow +\frac{1}{2}$	
30	$\sqrt{\left(\frac{p}{2}\right)^2 + q} + \frac{p}{2} = x.$

Few such examples of single quadratic equations have been preserved. Less rare are those of quadratic systems, at least among the texts we still have today. These systems often take one of the two following forms, either initially or after a series of transformations:

$$\begin{cases} x + y = p \\ x \cdot y = q \end{cases}$$

$$\begin{cases} x - y = p \\ x \cdot y = q. \end{cases}$$

Expressed in modern terms, the solution proceeds as follows. In the identity

$$\left(\frac{x+y}{2}\right)^2 - \left(\frac{x-y}{2}\right)^2 = xy,$$

two terms are known, and we can therefore calculate the term on the left side that is still unknown. Then, after taking its square root, we are able to determine the two unknown quantities x and y by using the identity previously encountered in solving linear systems:

$$x,\ y = \frac{x+y}{2} \pm \frac{x-y}{2}.$$

These are not the only identities known to the Mesopotamians, and other identities allow different systems to be solved. Thus,

$$\left(\frac{x+y}{2}\right)^2 + \left(\frac{x-y}{2}\right)^2 = \frac{x^2+y^2}{2}$$

is used to solve the system

$$\begin{cases} x^2 + y^2 = p \\ x \pm y = q, \end{cases}$$

and

$$\frac{x^2+y^2}{4} \pm \frac{x \cdot y}{2} = \left(\frac{x \pm y}{2}\right)^2$$

is used to solve

$$\begin{cases} x^2+y^2 = p \\ x \cdot y = q. \end{cases}$$

Example 3. Tablet 8862, Louvre Museum, Problem 1 [Appendix A.5]

Here the answer obtained is also verified at the end, but this time it is given immediately after the statement of the problem as well.

(α) *A rectangle. I multiplied the length by the width, I thus constructed an area. Then I added to the area that by which the length exceeds the width;* (*it gives*) 183. *Then I added the length to the width;* (*it gives*) 27. *What are the length, the width, and the area?*

27	183	*the sums*	
15	*the length*		
		180	*the area*
12	*the width.*		

You proceed thus. Add 27, *the sum of the length and the width, to* 183*;* (*it gives*) 210. *Add* 2 *to* 27*;* (*it gives*) 29. *Divide in two* 29*;* (*it gives*) $14+\frac{1}{2}$; *by* $14+\frac{1}{2}$, (*it gives*) $210+\frac{1}{4}$. *From* $210+\frac{1}{4}$ *you subtract* 210*;* (*it gives*) $\frac{1}{4}$, *the remainder.* $\frac{1}{4}$ *is the square of* $\frac{1}{2}$. *Add* $\frac{1}{2}$ *to the first* $14+\frac{1}{2}$; (*it gives*) 15, *the length. You subtract* $\frac{1}{2}$ *from the second* $14+\frac{1}{2}$; (*it gives*) 14, *the width. You subtract* 2, *which you added to* 27, *from* 14, *the width;* (*it gives*) 12, *the true width.*

I multiplied 15, *the length, by* 12, *the width.* 15 *by* 12 (*gives*) 180, *the area. By what does* 15, *the length, exceed* 12, *the width? It exceeds* (*it*) *by* 3. *Add* 3 *to* 180, *the area;* (*it gives*) 183, (*the sum of the excess of the length over the width and*) *the area.*

(β) We must solve the following system:

$$\begin{cases} xy + (x-y) = 183 \\ x + y = 27. \end{cases}$$

The calculations performed in the text are

$$\begin{aligned}
&27 + 183 = 210 \\
&27 + 2 = 29 \\
&29/2 = 14 + \frac{1}{2} \\
&\left(14 + \frac{1}{2}\right)^2 = 210 + \frac{1}{4} \\
&210 + \frac{1}{4} - 210 = \frac{1}{4} \\
&\sqrt{\frac{1}{4}} = \frac{1}{2} \\
&\left(14 + \frac{1}{2}\right) + \frac{1}{2} = 15 \qquad \text{"length"} \\
&\left(14 + \frac{1}{2}\right) - \frac{1}{2} = 14 \qquad \text{"width"} \\
&14 - 2 = 12 \qquad \text{"true width."}
\end{aligned}$$

The verification follows these calculations.

(γ) Interpretation: Again, we consider the system

$$\begin{cases} xy + (x - y) = 183 \\ x + y = 27. \end{cases}$$

Adding the two equations yields

$$xy + (x - y) + (x + y) = x(y + 2) = 210.$$

Now add 2 to both sides of the second equation to obtain

$$x + (y + 2) = 29.$$

Setting $y' = y + 2$ then yields the standard system already seen (page 11):

$$\begin{cases} x + y' = 29 \\ x \cdot y' = 210. \end{cases}$$

Indeed, the text calculates

$$\left(\frac{x + y'}{2}\right)^2 - xy' = \left(\frac{x - y'}{2}\right)^2$$

and then

$$\frac{x + y'}{2} \pm \frac{x - y'}{2} = x,\ y'.$$

The true width, as opposed to the "width" y', is then $y = y' - 2$. The text has thus performed what we would call a change of variable in order to reduce the proposed system to the standard form. Although nothing is explicitly said about this, the computations performed do not allow of any other interpretation.

Example 4. Tablet 13901, British Museum, Problem 18 [Appendix A.6]
(α) *I added the area of my three squares;* (*it gives*) 1400. *The side of one exceeds the side of the other by* 10.

You multiply by 1 *the* 10 *that exceeds;* (*it gives*) 10. *You multiply* (*it*) *by* 2*;* (*it gives*) 20. *You multiply* 20 *by* 20*;* (*it gives*) 400. *You multiply* 10 *by* 10*;* (*it gives*) 100. *You add* (*it*) *to* 400*;* (*it gives*) 500. *You subtract* (*it*) *from* 1400*;* (*it gives*) 900. *You multiply* (*it*) *by* 3, *the squares*[17]*; you write* 2700. *You add* 10 *and* 20*;* (*it gives*) 30. *You multiply* 30 *by* 30*;* (*it gives*) 900. *You add* (*it*) *to* 2700*;* (*it gives*) 3600. *It is the square of* 60. *You subtract* (*from it*) 30, *which you multiplied; you write* 30. *You multiply by* 30 *the inverse of* 3, *the squares,* (*which is*) $\frac{1}{3}$*;* (*it gives*) 10, *the side of the* (*first*) *square. You add* 10 *to* 10*;* (*it gives*) 20, *the side of the second square. You add* 10 *to* 20*;* (*it gives*) 30, *the side of the third square.*
(β) The calculations performed are

$$1 \cdot 10 = 10$$
$$2 \cdot 10 = 20$$
$$20 \cdot 20 = 400$$
$$10 \cdot 10 = 100$$
$$400 + 100 = 500$$
$$1400 - 500 = 900$$
$$3 \cdot 900 = 2700$$
$$10 + 20 = 30$$
$$30 \cdot 30 = 900$$
$$900 + 2700 = 3600$$
$$\sqrt{3600} = 60$$
$$60 - 30 = 30$$
$$30 \cdot \frac{1}{3} = 10, \text{ yielding the other values 20 and 30.}$$

(γ) Let us now examine the problem in our way. We are to solve

$$\begin{cases} x^2 + y^2 + z^2 = 1400 \\ x - y = 10 \\ y - z = 10. \end{cases}$$

Taking $y = z + 10$ and $x = y + 10 = z + 20$, the first equation becomes

$$(z + 20)^2 + (z + 10)^2 + z^2 = 1400.$$

Equivalently,

$$3z^2 + 2(20 + 10)z + (20^2 + 10^2) = 1400,$$

[17] That is, the *number* of squares (according to the statement, and also the coefficient of x^2 in the equation). This way of expressing a coefficient is common and not specifically Mesopotamian; it is found until the Renaissance (see pages 129–130, 135, 138, note 175).

or

$$3z^2 + 2 \cdot 30z + 500 = 1400.$$

Thus

$$3z^2 + 2 \cdot 30z = 900,$$

so

$$z = \frac{\sqrt{30^2 + 2700} - 30}{3} = \frac{60 - 30}{3} = 10.$$

Note that all numerical values occurring above are calculated successively in the text. Once again, however, there is no hint of the method followed.

Example 5. Tablet 13901, British Museum, Problem 9 [Appendix A.7]
(α) *I added the area of my two squares; (it gives)* 1300. *The side of one exceeds the side of the other by* 10.

You divide in two 1300*; you write* 650. *You divide in two* 10*; (it gives)* 5. *You multiply (it) by* 5*; (it gives)* 25. *You subtract (it) from* 650*; (it gives)* 625. *It is the square of* 25. *You write* 25 *twice. You add* 5*, which you multiplied, to the first* 25*; (it gives)* 30*, the side of the (first) square. You subtract* 5 *from the second* 25*; (it gives)* 20*, the side of the second square.*
(β) The calculations are

$$\begin{aligned}
&\frac{1300}{2} = 650\\
&\frac{10}{2} = 5\\
&5 \cdot 5 = 25\\
&650 - 25 = 625\\
&\sqrt{625} = 25\\
&25 + 5 = 30\\
&25 - 5 = 20.
\end{aligned}$$

(γ) The corresponding system is

$$\begin{cases} x^2 + y^2 = 1300 \\ x - y = 10. \end{cases}$$

We can therefore apply the identities seen above (page 11) to find x and y:

$$\sqrt{\frac{x^2 + y^2}{2} - \left(\frac{x - y}{2}\right)^2} \pm \frac{x - y}{2} = \frac{x + y}{2} \pm \frac{x - y}{2} = x,\ y.$$

This is indeed how the text proceeds.

Now, consider all these examples together. Had we not followed each text by its interpretation, we might have imagined that Mesopotamian algebra consisted of no more than a series of blind computations arriving at the correct answer by trial and error. After considering the interpretation, however, another picture emerges. The Mesopotamian student needed to memorize a small number of identities, and the art of solving these problems then consisted in transforming each problem into a standard form and calculating the solution, often using the half-sum or half-difference of the two unknowns at the end. Although this is certainly not the only solution method in use at the time, it is the most characteristic, which is why we have focused on it here. Moreover, as we will soon see, this method did not disappear with Mesopotamian mathematics.

Chapter 2

Algebra in Ancient Greece

2.1. Introduction

There are two very different sides of algebra in Ancient Greece. One is archaic, used mainly in schools, and recalls what we have seen in Mesopotamia. Although the problems of this first kind are more advanced than those found in Mesopotamia, the form is fundamentally the same: one finds sequences of unjustified calculations, and only the correctness of the answers suggests the existence of an underlying method. This is the kind we shall examine first, with a few second-degree problems (first part of Section 2.2). The second kind is Diophantine algebra. It is intended for higher mathematical education, and will look more familiar to us since there is a designated unknown, symbolism (even if rudimentary), and, above all, exposition that clearly shows the motivation for the given solution method. We shall speak about it at the end of this chapter (Section 2.3). Between these two aspects of second-degree algebra, we shall mention a few linear problems (second part of Section 2.2), starting with simple (school) examples and ending with Archimedes' famous cattle problem.

Remark.

- From the name "Diophantine algebra," it should not be inferred that the evolved form of algebra originated with Diophantus; he is simply our only extant Greek source for this mathematics, and he must surely not have been the only one to treat the subject, albeit certainly the most famous one.

The number system used in Greece is very distinctive. An ancient system, appearing in the fifth century B.C. and evidently developed after the written language, used the first letters of the names of the powers of 10 as symbols (Δ, H, X meaning 10, 100, 1000 respectively, while I represented the unit). These were then juxtaposed as many times as necessary; thus XHHΔΔΔIIII meant 1234. (This was simplified somewhat by combining one of these symbols with Π, the first letter of "five," to denote five times the power in question.) This system was fairly rapidly replaced by the more convenient classical system, itself clearly established as a rational construction by an individual or group of people. The idea of expressing numerals by letters was retained: the twenty-four letters of the classical Greek alphabet, augmented by an archaic letter and two regional letters borrowed from Phoenician, formed a set of twenty-seven symbols used to denote the nine units, nine multiples of ten, and nine multiples of one hundred. A line placed above the expression for a number allowed one to immediately distinguish symbols denoting numbers from words. For the nine multiples of one thousand, a comma was juxtaposed with the symbols used for units (see Figure 1).[18] This system allowed one to denote any number from 1 to 9999 without requiring a symbol for zero. For higher numbers, the same thirty-six symbols were used, with tens of thousands being separated by the symbol M^{Y} repeated as necessary, and the first group being preceded by the symbol M^{o} (or sometimes only a dot). Thus 120 378 449 500, read as 1203 7844 9500, would be written as $\mathrm{M}^{\mathrm{Y}}\mathrm{M}^{\mathrm{Y}}\overline{,\alpha\sigma\gamma}\,\mathrm{M}^{\mathrm{Y}}\overline{,\zeta\omega\mu\delta}\,\mathrm{M}^{\mathrm{o}}\overline{,\vartheta\varphi}$.[19]

$\overline{\alpha}$	$\overline{\beta}$	$\overline{\gamma}$	$\overline{\delta}$	$\overline{\varepsilon}$	$\overline{\varsigma}$	$\overline{\zeta}$	$\overline{\eta}$	$\overline{\vartheta}$
1	2	3	4	5	6	7	8	9
$\overline{\iota}$	$\overline{\kappa}$	$\overline{\lambda}$	$\overline{\mu}$	$\overline{\nu}$	$\overline{\xi}$	$\overline{o}$	$\overline{\pi}$	$\overline{\text{ϙ}}$
10	20	30	40	50	60	70	80	90
$\overline{\rho}$	$\overline{\sigma}$	$\overline{\tau}$	$\overline{\upsilon}$	$\overline{\varphi}$	$\overline{\chi}$	$\overline{\psi}$	$\overline{\omega}$	$\overline{\text{ϡ}}$
100	200	300	400	500	600	700	800	900
$\overline{,\alpha}$	$\overline{,\beta}$	$\overline{,\gamma}$	$\overline{,\delta}$	$\overline{,\varepsilon}$	$\overline{,\varsigma}$	$\overline{,\zeta}$	$\overline{,\eta}$	$\overline{,\vartheta}$
1000	2000	3000	4000	5000	6000	7000	8000	9000

FIGURE 1

Remark.

- Following the modern convention, we have used minuscule Greek letters. This is anachronistic since only capitals were used throughout antiquity (the introduction of minuscules dates from around A.D. 800, and is therefore Byzantine).

[18] The archaic letter *digamma* (for 6) retained its place in the old alphabet; the two supplementary letters were put at the end of the tens and of the hundreds. These three letters may vary in form (we use a different form for 6 in Appendix B).

[19] On number systems in general, see K. Menninger, *Zahlwort und Ziffer*, Breslau 1934; English translation from the second German edition (Göttingen 1957–58) as *Number Words and Number Symbols*, Cambridge (Mass.) 1969 (reprinted).

2.2. Common Algebra

For second-degree problems, by "common algebra" we mean those of the archaic kind, as practiced in schools. It survives today in small groups of problems with neither title nor author, found either on fragments of papyrus or incorporated into larger works. We will first consider these quadratic problems, as their treatment is quite close to what we have already seen in Mesopotamia.

Problems About Right Triangles

The examples we will consider use identities similar to the ones we already know from Chapter 1 (see page 11). These are

(1) $x, y = \frac{1}{2}\left[(x+y) \pm (x-y)\right]$
(2) $x^2 + y^2 \pm 2xy = (x \pm y)^2$,

and, as inferred from the latter,

(3) $(x+y)^2 - (x-y)^2 = 4xy$,
(4) $(x+y)^2 + (x-y)^2 = 2(x^2+y^2)$.

As these problems concern right triangles, whose sides therefore satisfy the relation $u^2 + v^2 = w^2$ and which have area $A = \frac{1}{2}uv$, the last three identities become

(2′) $w^2 \pm 4A = (u \pm v)^2$
(3′) $(u+v)^2 - (u-v)^2 = 8A$
(4′) $(u+v)^2 + (u-v)^2 = 2w^2$.

We will follow the Greek convention of supposing that the base u of the triangle is greater than the height v (Figure 2).

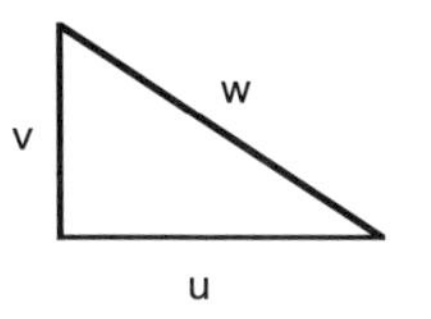

FIGURE 2

Example 1. Greek papyrus 259, Geneva Library [Appendix B.1]

This papyrus, which was probably written in the second century, lacks its left and right parts, and today only three problems are preserved. The first is a simple application of the Pythagorean theorem, while the other two are solved using the above identities. These two are the following[20].

[20] Publication of this papyrus in J. Sesiano, "Sur le Papyrus graecus genevensis 259," *Museum Helveticum*, 56 (1999), pp. 26–32. One can also find the text and analysis of these and similar problems in J. Sesiano, "An early form of Greek algebra," *Centaurus*, 40 (1998), pp. 276–302. As is the case for Mesopotamian texts, words in parentheses have been added by us to clarify meaning.

There is a right-angled triangle of which the height and the hypotenuse are together 8 *feet and the base is* 4 *feet; we shall look for its height and its hypotenuse separately.*

We shall find (them) as follows. The[21] 4 *by itself, there results* 16. *Divide (it) by the* 8, *there results* 2. *Subtract the* 2 *from the* 8, *there remains* 6. *Half of it is* 3. *The height will be* 3. *Next, subtract the* 3 *from the* 8, *there remains* 5. *Hence the hypotenuse will be* 5 *feet.*

There is a right-angled triangle of which the height and the base are together 17 *feet and the hypotenuse is* 13 *feet; to find the height and the base separately.*

We shall find (them) as follows. The 13 *by itself (gives)* 169. *And the* 17 *by itself, there results* 289. *Take twice the* 169, *there results* 338. *And subtract the* 289 *from the* 338*; there remains* 49. *Its side*[22] *is* 7. *Subtract this from the* 17, *there remains* 10. *Its half is* 5. *The height will be* 5. *Subtract this from the* 17, *there remains* 12. *Hence the base will be* 12 *feet.*

Thus, for the first problem, suppose we are given the base $u = l$ and the sum of the hypotenuse and the height $w + v = k$. As $u^2 = w^2 - v^2 = (w+v)(w-v)$, we have

$$w - v = \frac{u^2}{w+v} = \frac{l^2}{k}.$$

We thus know $w - v$, and $w + v = k$, so we can calculate the height:

$$v = \frac{1}{2}\left[(w+v) - (w-v)\right] = \frac{1}{2}\left[k - \frac{l^2}{k}\right].$$

Then the hypotenuse is

$$w = (w+v) - v = k - v.$$

This is exactly how v and w are determined in the text, by means of the given numerical values $w + v = k = 8$ and $u = l = 4$.

In the second problem, we are given $u + v = k$ and $w = l$. As $u - v = \sqrt{2w^2 - (u+v)^2} = \sqrt{2l^2 - k^2}$ by identity $(4')$, we can calculate

$$v = \frac{1}{2}\left[(u+v) - (u-v)\right] = \frac{1}{2}\left[k - \sqrt{2l^2 - k^2}\right],$$

and then

$$u = (u+v) - v = k - v.$$

This is indeed how v and u are determined; with $k = 17$, $l = 13$, we have the successive computations of l^2, k^2, $2l^2$, $2l^2 - k^2$, $\sqrt{2l^2-k^2}$, $k - \sqrt{2l^2-k^2}$, $\frac{1}{2}[k - \sqrt{2l^2-k^2}] = v$, whence u.

[21]This use of the article in the Greek text indicates that the number in question has already been encountered, either as a given value or as the result of a calculation.

[22]That is, its root. The Greek word for "side" is commonly used to mean this.

Example 2. Problem from a Roman compendium on surveying [Appendix B.2]

While most problems written by Roman surveyors are simple and practical, some, the Latin text of which suggests a Greek origin (see note 127), are of a different nature. One example is the following problem[23].

For a right-angled triangle in which the measure in feet of the hypotenuse is 25 *feet (and) the area* 150 *(square) feet*[24]*, tell the height and the base separately.*

It is sought as follows. I always[25] *multiply the hypotenuse by itself, there results* 625. *I add to this quantity* 4 *areas, which makes* 600 *feet; both together, there results* 1225 *feet. I take the side of this, there results* 35. *(This is the sum of the two straight lines.) Next, in order to find the difference of the two straight lines: I shall make the measure of the hypotenuse by itself, there results* 625.[26] *I subtract from this* 4 *areas, there results* 25 *feet; I take the side of this, there results* 5*; it will be the difference. I always add this to the two added (straight lines), that is, to* 35*, there results* 40 *feet. I always take half of this, there results* 20 *feet. It will be the base of the triangle. If I subtract from* 20 *the difference, that is,* 5 *feet, there remains* 15 *feet. It will be the height of this triangle.*

Knowing $w = k$ and the area $\frac{1}{2}uv = A = l$, we wish to find u and v. By identity $(2')$, namely

$$w^2 \pm 4A = (u \pm v)^2,$$

we have

$$u + v = \sqrt{w^2 + 4A} = \sqrt{k^2 + 4l}$$
$$u - v = \sqrt{w^2 - 4A} = \sqrt{k^2 - 4l}.$$

It follows that

$$u = \frac{1}{2}\left[\sqrt{k^2 + 4l} + \sqrt{k^2 - 4l}\right],$$

whence

$$v = u - (u - v) = u - \sqrt{k^2 - 4l}.$$

Indeed, this is how the text calculates u and v from the known values $k = 25$ and $l = 150$.

[23]See F. Blume, K. Lachmann, and A. Rudorff, *Die Schriften der römischen Feldmesser* (2 vol.), Berlin 1848–52 (reprint: Hildesheim 1967), I, pp. 297–98, or (a better text) N. Bubnov, *Gerberti postea Silvestri II papae Opera mathematica*, Berlin 1899, pp. 511–12 (or the second reference mentioned in note 20).

[24]"foot" is used indistinctly for length and area measurement.

[25]In these texts (see also the following example), "always" means that the rule is general and independent of the values given in the problem.

[26]Computed again since this is another part of the problem.

Example 3. A group of four indeterminate problems, of a higher level of difficulty than the ones above, form part of a collection of writings attributed (not always correctly) to Heron of Alexandria (ca. A.D. 65).[27] We must determine the sides of a right triangle from the given (integral) sum of perimeter and area. We will first solve this question as above, limiting ourselves to using only identities, and then consider the actual calculations.

Let

$$P + A = k,$$

with k a given integer. Then

$$(u + v + w) + A = k.$$

Since, by identity ($2'$),

$$(u + v)^2 - w^2 = 4A,$$

we deduce that

$$A = \left(\frac{u + v + w}{2}\right)\left(\frac{u + v - w}{2}\right). \tag{*}$$

It follows that

$$\begin{aligned} P + A &= u + v + w + \left(\frac{u + v + w}{2}\right)\left(\frac{u + v - w}{2}\right) \\ &= \left(\frac{u + v + w}{2}\right)\left(\frac{u + v - w}{2} + 2\right) \\ &= k. \end{aligned}$$

Write

$$\begin{aligned} \frac{u + v + w}{2} &= p \\ \frac{u + v - w}{2} &= q - 2, \end{aligned}$$

so that k is the product of p and q. By adding and subtracting these two expressions (thus by applying the usual half-sum and half-difference identities), we obtain the two relations

$$u + v = p + q - 2 \tag{α}$$
$$w = p - (q - 2). \tag{β}$$

On the other hand, by expression (*),

$$A = p(q - 2).$$

[27] See *Heronis Alexandrini Opera quae supersunt omnia*, Vol. IV (edd. G. Schmidt and J. Heiberg), Leipzig 1912 (reprint: Stuttgart 1976), pp. 422,10–427. The text of these problems is reproduced, with some minor alterations, as an appendix in the second study mentioned in note 20.

Thus, by identity (3′),

$$(u-v)^2 = (u+v)^2 - 8A$$

$$(\gamma) \qquad = (p+q-2)^2 - 8p(q-2).$$

Hence, identity (1), applied to (α) and (γ), yields

$$\begin{aligned} u, v &= \frac{1}{2}\left[(u+v) \pm (u-v)\right] \\ &= \frac{1}{2}\left[(p+q-2) \pm \sqrt{(p+q-2)^2 - 8p(q-2)}\right], \end{aligned}$$

while, as seen in (β),

$$w = p - (q-2).$$

As we are considering only right triangles with rational sides, we must determine which of the possible representations of k as the product of two factors will yield a square radicand. (Note that if these two factors are integers, the same will be true for the legs of the triangles, since the two terms in square brackets have the same parity.)

The text of the first problem in this group will show that the above formula is just applied, without the slightest explanation as to its origin. The author simply indicates which is the appropriate choice among the various integral decompositions of k, and then calculates successively u and v by the above formula, and then $w = p - (q-2)$. Finally, he finds the values of $A = \frac{1}{2}uv$ and $u + v + w$ and checks that their sum is indeed equal to k. Here is the translation of this problem [Appendix B.3].

The area together with the perimeter of a right-angled triangle is 280 *feet; to separate the sides and to find the area.*

I proceed as follows. Always search for the numbers making up (280). *Now* 280 *is made up by twice* 140, 4 (*times*) 70, 5 (*times*) 56, 7 (*times*) 40, 8 (*times*) 35, 10 (*times*) 28, 14 (*times*) 20. *We have observed that the* 8 *and* 35 *will satisfy the given requirement.* [*The* $\frac{1}{8}$ *of the* 280 *is* 35 *feet.*][28] *Always take two away from the* 8, *there remains* 6 *feet. Then the* 35 *and the* 6 *together, there results* 41 *feet. Multiply this by itself, there results* 1681 *feet. The* 35 *by the* 6, *there results* 210 *feet. Always multiply this by the* 8, *there results* 1680 *feet. Subtract this from the* 1681, *there remains* 1. *Its square side is* 1. *Now take the* 41 *and subtract* 1 *unit; there remains* 40. *Its half is* 20. *This is the height,* 20 *feet. And take again the* 41 *and add* 1; *there results* 42 *feet. Its half is* 21 *feet. Let the base be* 21 *feet. And take the* 35 *and subtract the* 6; *there remains* 29 *feet. Let the hypotenuse be* 29 *feet. Now take the height and multiply* (*it*) *by the base; there results* 420 *feet. Its half is* 210 *feet. Such is the area. And the three enclosing sides make* 70 *feet. Add* (*it*) *together with the area, there results* 280 *feet.*

[28]This must be a former reader's remark later incorporated into the text. He thought that fulfilling the given requirement meant that the product $8 \cdot 35$ makes up 280 (whereas it actually has to do with the rationality of the solution).

The possible factorizations of $k = 280$ into two integers are, indeed, $2 \cdot 140 = 4 \cdot 70 = 5 \cdot 56 = 7 \cdot 40 = 8 \cdot 35 = 10 \cdot 28 = 14 \cdot 20$, and only the pair $p = 35$, $q = 8$ is acceptable, that is, gives rational sides for the triangle. For, as calculated in the text (with the numbers independent of the data being given in italics)

$$\begin{aligned} u,\, v &= \frac{1}{2}\left[(35 + 8 - \mathit{2}) \pm \sqrt{(35 + 8 - \mathit{2})^2 - \mathit{8} \cdot 35(8 - \mathit{2})}\right] \\ &= \frac{1}{2}\left[41 \pm \sqrt{1681 - 1680}\right], \end{aligned}$$

whence $u = 21$, $v = 20$, $w = 35 - 6 = 29$.

The three subsequent problems of this group are treated in exactly the same manner: the factorizations are listed, the appropriate one is chosen and the formula is then applied. In the first of these problems, $k = 270 = 2 \cdot 135 = 3 \cdot 90 = 5 \cdot 54 = 6 \cdot 45 = 9 \cdot 30 = 10 \cdot 27 = 15 \cdot 18$, and only the pair $p = 45$, $q = 6$ gives a rational solution. Then $p + q - 2 = 49$, and $q - 2 = 4$, so

$$u,\, v = \frac{1}{2}\left[49 \pm \sqrt{2401 - 1440}\right] = \frac{1}{2}\left[49 \pm 31\right],$$

and $u = 40$, $v = 9$, $w = 41$.

In the next problem, $k = 100 = 2 \cdot 50 = 4 \cdot 25 = 5 \cdot 20 = 10 \cdot 10$; among these decompositions, only $p = 20$, $q = 5$ can be chosen. Then $p+q-2 = 23$, $q - 2 = 3$, and

$$u,\, v = \frac{1}{2}\left[23 \pm \sqrt{529 - 480}\right] = \frac{1}{2}\left[23 \pm 7\right],$$

so $u = 15$, $v = 8$, $w = 17$.

Finally, in the last problem, $k = 90 = 2{\cdot}45 = 3{\cdot}30 = 5{\cdot}18 = 6{\cdot}15 = 9{\cdot}10$, and the author considers the pair $p = 18$, $q = 5$. Then $p + q - 2 = 21$ and $q - 2 = 3$, so

$$u,\, v = \frac{1}{2}\left[21 \pm \sqrt{441 - 432}\right] = \frac{1}{2}\left[21 \pm 3\right],$$

whence $u = 12$, $v = 9$, $w = 15$.

However, in this case there is a second possible decomposition, not considered by the text: $p = 3$, $q = 30$. Then $p + q - 2 = 31$ and $q - 2 = 28$, yielding

$$u,\, v = \frac{1}{2}\left[31 \pm \sqrt{961 - 672}\right] = \frac{1}{2}\left[31 \pm 17\right],$$

so that $u = 24$, $v = 7$. This other possibility was certainly intentionally disregarded: applying the formula for w gives the negative value $w = p - (q-2) = -25$. Observe, though, that $24^2 + 7^2 = 25^2$ would be an acceptable solution. We will see that in Diophantine algebra as well, a negative solution is not accepted, even when the quantity appears only to the square in the problem to be solved. As we shall see at the end of this chapter, Euler takes a different attitude in similar problems. But this was fifteen centuries later, and the attitude towards negative numbers had changed.

These Greek examples of common algebra thus remind us of the Mesopotamian way of treating quadratic problems by means of identities. Whether this similarity was originally the result of an influence and became a traditional approach or whether it arose independently, as the natural way to treat such problems with no algebra available, must remain an open question.

Linear Problems

Book XIV of a collection of epigrams, now known as *Greek Anthology* (*Anthologia Graeca* or *Anthologia Palatina*), consists of various puzzles and little arithmetic problems. Although they do not include any algebra (only the statements are given), it is worth mentioning them here as they belong to the area of recreational mathematics, a domain that was to grow considerably during the Middle Ages to the point where it became a standard component of works on algebra (Chapter 3, page 64).

Epigram 126 reports the poem allegedly inscribed on the tombstone of Diophantus. Its solution reveals the age attained by Diophantus (84) as well as some stages in his life. We learn that Diophantus's childhood lasted a sixth of his life, that after a twelfth of his life he could shave for the first time (an important event in antiquity, since it meant the passage to adult age); that after a seventh he married; five years later, he fathered a son, who died upon reaching half the years that his father would live. Finally, during the four remaining years of his life, Diophantus attempted to find peace and solace in mathematics. As for the other epigrams, the solution is not given, but it is easily inferred from $\frac{1}{6}x + \frac{1}{12}x + \frac{1}{7}x + 5 + \frac{1}{2}x + 4 = x$, whereby the stages of his life are found (14, 21, 33, 38, 80).

Another problem of the *Anthologia* (number 132) shows that so-called cistern problems already served to trouble school children. It is about the cyclops Polyphemus, one of the numerous obstacles Ulysses found on his way back to Ithaca at the end of the Trojan War. From a statue of Polyphemus, we learn, water flowed through three openings. The stream emerging from the hand alone would fill the basin in three days; the one pouring from his single eye would do this in one day; and the one running from the mouth would need only two-fifths of a day. The question is then how much time would be required to fill the basin if water flowed from the three openings simultaneously. As for the problem above, the reader is left to find the solution ($\frac{6}{23}$ of a day).

We find on a papyrus, probably dating from the second century, three systems of linear equations which are solved using a designated unknown.

One of them is[29]

$$\begin{cases} u+v+w+t=9900 \\ v-u=\frac{1}{7}u \\ w-(u+v)=300 \\ t-(u+v+w)=300. \end{cases}$$

This is solved as follows. We will express each of the four desired quantities in terms of an auxiliary unknown x. Set $u=7x$. By the second equation, then $v=8x$, so $u+v=15x$. By the third equation, $w=u+v+300=15x+300$, so $u+v+w=30x+300$. By the fourth equation, $t=u+v+w+300=30x+600$. Then, by adding these last two equalities and using the first equation, we obtain $u+v+w+t=60x+900=9900$, so that $x=150$, and $u=1050$, $v=1200$, $w=2550$, and $t=5100$. It is important to observe that the particular symbol used to denote the unknown x in this problem is exactly the same as used by Diophantus.

We know that the Greeks solved systems of linear equations that were significantly more complicated than those that have reached us on a few papyri. Indeed, Iamblichus, one of our late sources of information on the Pythagoreans, shows us how one can apply a rule called, rather affectedly, *epanthema* (ἐπάνθημα, bloom), which dates back to Thymaridas, a mathematician living before 300 B.C.

Consider the system (written in modern notation; as usual, the text describes it verbally)

$$\begin{cases} x+x_1+x_2+\cdots+x_{n-1}=S \\ x+x_1=a_1 \\ x+x_2=a_2 \\ \dots \\ x+x_{n-1}=a_{n-1} \end{cases}$$

consisting of n equations with n unknowns x, x_i and n given values S, a_i. According to Thymaridas's rule, the main unknown x (from which the values of the remaining unknowns are inferred) is determined by

$$x=\frac{(a_1+a_2+\ldots+a_{n-1})-S}{n-2},$$

as adding the last $n-1$ equations yields

$$(n-2)x+S=a_1+a_2+\cdots+a_{n-1}.$$

[29]L. Karpinski and F. Robbins, "Michigan Papyrus 620," *Science*, 70 (1929), pp. 311–314. Greek text in J. Winter *Michigan papyri*, Vol. III, Ann Arbor 1936, pp. 29–30 (translation and commentary pp. 30–34).

Iamblichus shows how to apply this rule to other linear systems. Following his example, consider the system (again written in modern notation)

$$\begin{cases} x_1 + x_2 = a(x_3 + x_4) \\ x_1 + x_3 = b(x_2 + x_4) \\ x_1 + x_4 = c(x_2 + x_3), \end{cases}$$

where a, b, and c are given quantities. By adding to each equation the factor appearing in parentheses, we obtain

$$\begin{cases} x_1 + x_2 + x_3 + x_4 = (a+1)(x_3 + x_4) \\ x_1 + x_2 + x_3 + x_4 = (b+1)(x_2 + x_4) \\ x_1 + x_2 + x_3 + x_4 = (c+1)(x_2 + x_3). \end{cases}$$

Let S be the sum of the four unknowns. Then the first of these equations becomes

$$x_3 + x_4 = \frac{1}{a+1} S.$$

Substituting the right side for $x_3 + x_4$ in the first given equation yields

$$x_1 + x_2 = \frac{a}{a+1} S,$$

while proceeding similarly for the other two equations yields

$$x_1 + x_3 = \frac{b}{b+1} S$$

$$x_1 + x_4 = \frac{c}{c+1} S.$$

This brings us to the situation of Thymaridas's rule, which we can apply to obtain

$$x_1 = \frac{S\left(\frac{a}{a+1} + \frac{b}{b+1} + \frac{c}{c+1}\right) - S}{2}.$$

As the problem is indeterminate, we can assign a value to S; a suitable choice will yield integral solutions.

The *Odyssey* was not only the source of inspiration for the above cistern problem. Another episode from the return of Ulysses to Ithaca gave Archimedes (ca. 287 B.C.–212 B.C.) the idea for a problem that he proposed as a challenge to his colleagues in Alexandria.[30] The statement alone, in verse, is contained in a letter he sent to Eratosthenes. The goddess Circe had predicted that on his way to Ithaca, Ulysses would stop on the island of Thrinacia (Sicily), where the cattle of Helios, the sun god, grazed. Furthermore, she had warned him that if any of his crew were to harm in any manner the cattle, all would perish in a shipwreck and Ulysses alone would finally succeed, under miserable conditions, in reaching Ithaca. Failing to

[30] Archimedes himself lived in Syracusa. At that time, Syracusa (in Sicily) and Alexandria (in Egypt) were Greek cities.

heed the warnings of Ulysses, the crew took advantage of Ulysses's sleep to allay their hunger, thereby sealing their tragic fate.

According to the *Odyssey*, the number of cattle of the Sun is significant, but not too large. Archimedes made it a huge number, which the Alexandrian mathematicians had to determine from his clues. We are told that the cattle were divided into four herds of bulls and cows, with (say) x_1 and y_1 respectively having white coats, x_2 and y_2 having black coats, x_3 and y_3 having spotted coats, and x_4 and y_4 having brown coats, and seven relations between these quantities being given (thus one less than the quantity of unknowns). According to Archimedes, he who could use these relations to determine the unknown quantities could not be considered to be ignorant or incompetent in the realm of science, yet neither would he be worthy of being called a scholar. But he would reach perfection of knowledge if he succeeded in determining all these quantities by taking into account two additional conditions: the black and white bulls together can be arranged to form a square, while the brown and spotted bulls together can form a triangle with a single bull at its apex.

The seven given relations between the various types of cattle can be translated mathematically into the following equations:

$$
\begin{aligned}
&(1)\ x_1 = \left(\tfrac{1}{2} + \tfrac{1}{3}\right) x_2 + x_4 = \tfrac{5}{6} x_2 + x_4 \\
&(2)\ x_2 = \left(\tfrac{1}{4} + \tfrac{1}{5}\right) x_3 + x_4 = \tfrac{9}{20} x_3 + x_4 \\
&(3)\ x_3 = \left(\tfrac{1}{6} + \tfrac{1}{7}\right) x_1 + x_4 = \tfrac{13}{42} x_1 + x_4 \\
&(4)\ y_1 = \left(\tfrac{1}{3} + \tfrac{1}{4}\right) (x_2 + y_2) = \tfrac{7}{12} (x_2 + y_2) \\
&(5)\ y_2 = \left(\tfrac{1}{4} + \tfrac{1}{5}\right) (x_3 + y_3) = \tfrac{9}{20} (x_3 + y_3) \\
&(6)\ y_3 = \left(\tfrac{1}{5} + \tfrac{1}{6}\right) (x_4 + y_4) = \tfrac{11}{30} (x_4 + y_4) \\
&(7)\ y_4 = \left(\tfrac{1}{6} + \tfrac{1}{7}\right) (x_1 + y_1) = \tfrac{13}{42} (x_1 + y_1).
\end{aligned}
$$

Therefore, from the first three,

$$
\begin{aligned}
&(1')\ 6x_1 - 5x_2 = 6x_4 \\
&(2')\ 20x_2 - 9x_3 = 20x_4 \\
&(3')\ 42x_3 - 13x_1 = 42x_4.
\end{aligned}
$$

Eliminating x_2 by combining $(1')$ and $(2')$ and then x_3 by combining the resulting equation and $(3')$ yields

$$297x_1 = 742x_4,$$

whence

$$
\begin{aligned}
x_1 &= \frac{742}{297} x_4 \\
x_2 &= \frac{178}{99} x_4 \\
x_3 &= \frac{1580}{891} x_4.
\end{aligned}
$$

Since the system is indeterminate, we can choose x_4; as the fractions are reduced to lowest terms, we will take $x_4 = 891k$, with k a positive integer, to obtain a solution in integers. Then $x_3 = 1580k$, $x_2 = 1602k$, and $x_1 = 2226k$. Introducing these expressions into the four equations for the cows yields

$$
\begin{aligned}
&(4')\ 12y_1 - 7y_2 = 7x_2 = 11\,214 \cdot k\\
&(5')\ 20y_2 - 9y_3 = 9x_3 = 14\,220 \cdot k\\
&(6')\ 30y_3 - 11y_4 = 11x_4 = 9801 \cdot k\\
&(7')\ 42y_4 - 13y_1 = 13x_1 = 28\,938 \cdot k.
\end{aligned}
$$

Then, by successively eliminating y_2, y_3, and y_4, we obtain

$$
\begin{aligned}
y_1 &= \frac{7\,206\,360}{4657} \cdot k\\
y_2 &= \frac{4\,893\,246}{4657} \cdot k\\
y_3 &= \frac{3\,515\,820}{4657} \cdot k\\
y_4 &= \frac{5\,439\,213}{4657} \cdot k.
\end{aligned}
$$

As the denominator is prime and does not divide the numerators, we set $k = 4657 \cdot l$, with l a positive integer. Then

$$
\begin{aligned}
x_1 &= 10\,366\,482 \cdot l\\
x_2 &= 7\,460\,514 \cdot l\\
x_3 &= 7\,358\,060 \cdot l\\
x_4 &= 4\,149\,387 \cdot l,
\end{aligned}
$$

while, clearly,

$$
\begin{aligned}
y_1 &= 7\,206\,360 \cdot l\\
y_2 &= 4\,893\,246 \cdot l\\
y_3 &= 3\,515\,820 \cdot l\\
y_4 &= 5\,439\,213 \cdot l.
\end{aligned}
$$

The total is then $50\,389\,082 \cdot l$ animals.

So much for the part accessible to the average mathematician. Not surprisingly, since it is a fairly simple algebraic manipulation, we find this smallest integral solution calculated in Greek sources. But we must now also take into account the last two conditions. According to the first, the sum of x_1 and x_2 must be a square. As

$$x_1 + x_2 = 17\,826\,996 \cdot l = 2^2 \cdot 3 \cdot 11 \cdot 29 \cdot 4657 \cdot l,$$

we will obtain a square by taking $l = 3 \cdot 11 \cdot 29 \cdot 4657 \cdot m^2$, yielding

$$\begin{aligned}
x_1 &= 46\,200\,808\,287\,018 \cdot m^2\\
x_2 &= 33\,249\,638\,308\,986 \cdot m^2\\
x_3 &= 32\,793\,026\,546\,940 \cdot m^2\\
x_4 &= 18\,492\,776\,362\,863 \cdot m^2\\
y_1 &= 32\,116\,937\,723\,640 \cdot m^2\\
y_2 &= 21\,807\,969\,217\,254 \cdot m^2\\
y_3 &= 15\,669\,127\,269\,180 \cdot m^2\\
y_4 &= 24\,241\,207\,098\,537 \cdot m^2,
\end{aligned}$$

for a total of $224\,571\,490\,814\,418 \cdot m^2$ animals.

Next, according to the second condition, $x_3 + x_4$ must form a triangle. The concept of so-called "triangular numbers" was as familiar to the Greeks as that of square numbers. The ancient Pythagoreans, in the fifth century B.C., called *triangular* a number of which the units, represented as points, could be arranged in an equilateral triangle (Figure 3). Such a triangle of order n, T_n, is thus made up of the series $1 + 2 + 3 + \cdots + n$, and it had been observed that assembling such triangles could be used to display various properties. For instance, joining two triangles of the same order (Figure 4) forms the rectangle $2T_n = n(n+1)$, which yields an expression for the sum of consecutive positive integers beginning with 1 when the last term is known. Similarly, placing eight identical triangles around a point (Figure 5) forms a square, as

$$8 \cdot T_n + 1 = 8 \cdot \frac{n(n+1)}{2} + 1 = 4n(n+1) + 1 = (2n+1)^2.$$

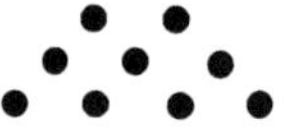

FIGURE 3

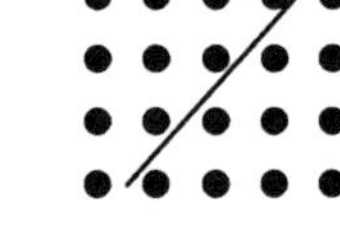

FIGURE 4

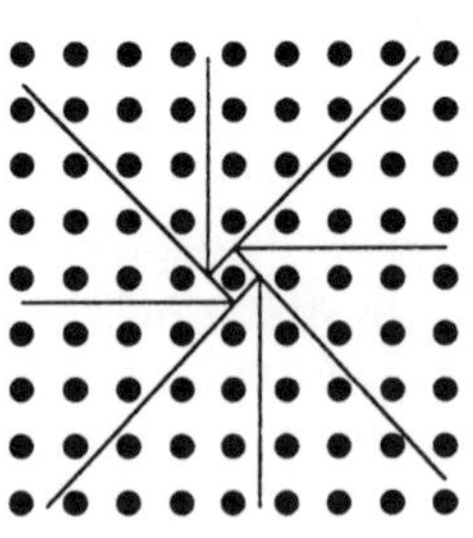

FIGURE 5

In the case of Archimedes' problem, we must have

$$x_3 + x_4 = 51\,285\,802\,909\,803 \cdot m^2 = T_n = \frac{n(n+1)}{2}.$$

Factorizing the number and applying the above property then yields

$$2^3 \cdot 3 \cdot 7 \cdot 11 \cdot 29 \cdot 353 \cdot 4657^2 \cdot m^2 + 1 = (2n+1)^2.$$

Next, writing $t = 2n + 1$ and $u = 2 \cdot 4657 \cdot m$ yields

$$t^2 - 2 \cdot 3 \cdot 7 \cdot 11 \cdot 29 \cdot 353 \cdot u^2 = t^2 - 4\,729\,494 \cdot u^2 = 1,$$

an equation to be solved in the natural numbers. Its minimal solution, as calculated by A. Amthor in 1880, is

$$t_1 = 109\,931\,986\,732\,829\,734\,979\,866\,232\,821\,433\,543\,901\,088\,049$$
$$u_1 = 50\,549\,485\,234\,315\,033\,074\,477\,819\,735\,540\,408\,986\,340.$$

However, this solution does not meet our requirements, because the solution u must be divisible by $2 \cdot 4657$ for m to be an integer, and this is not the case for u_1. Instead, this condition will be satisfied by u_{2329}. The number of cattle of the Sun may thus be determined—at least in theory, since it is a 206 549-digit number![31]

2.3. Diophantine Algebra

Diophantus and his *Arithmetica*

Diophantus is one of the scholars of antiquity whose name is most frequently mentioned by mathematicians of today, particularly in the context of number theory and algebraic geometry. Few of them, however, have much idea of his work itself. Diophantus owes his reputation—and the liberal use of his name—to his influence during the seventeenth century, particularly when Fermat (1601–1665) took the *Arithmetica* as the starting point for his studies in number theory.

Of Diophantus, we know only that he lived in Alexandria, the center of Greek scientific activity throughout antiquity. He is generally placed around A.D. 250, as on the one hand Diophantus is cited by Theon of Alexandria, who lived around A.D. 350, and on the other hand none of the mathematicians of the first two centuries mention him, even though some of them had numerous occasions to do so. Other particulars of his life are not known; one should not trust those given in the little poem considered above (page 25).

To Diophantus is attributed a text on summing arithmetic series, of which only a small fragment remains today. But his reputation stems uniquely from his *Arithmetica*, a collection of problems divided into thirteen books

[31]The problem of the cattle of the Sun, first published by G. E. Lessing in 1773, was solved by A. Amthor, "Das Problema bovinum des Archimedes," *Zeitschrift für Mathematik und Physik, Histor.-literar. Abtlg*, XXV (1880), pp. 156–171. The last two conditions lead to solving a so-called "Fermat" or (by misattribution) "Pell" equation, which requires the use of continued fractions.

(βιβλία), or chapters, of which ten have been preserved. Six are known in the original Greek text, although augmented by some problems added later by glossators. These books, which were studied in Europe from the late sixteenth century (the Middle Ages knew nothing of the *Arithmetica*), are the origin of early modern number theory. The other four books are preserved in Arabic translation, in a manuscript located by F. Sezgin in 1968 in the Astan Qods Library in Mashhad (Iran), before a catalog of that library published in 1971 made its existence widely known. The study of these four books revealed that Diophantus's text as translated into Arabic in the ninth century differed in form from the known Greek version: the Arabic text appears to be much wordier, coming from a reworked Greek text that may have been the commentary that Greek sources attribute to Hypatia, the daughter of Theon of Alexandria.

Thus today the extant *Arithmetica* comprises Books I to III (in Greek), IV to VII (in Arabic; the manuscript does not contain the three preceding books, known to have been translated as well), and finally "IV" to "VI" (in Greek), which were misnumbered by the Byzantines during the Middle Ages and may have originally been Books VIII to X. While Book I presents elementary problems singularly reminiscent of the problems we have seen above, Books II and III (particularly II) explain and apply fundamental methods, which Books IV to VII use extensively, thereby demonstrating their diverse applications. Finally, Books "IV" to "VI" solve problems of a higher level of difficulty, mainly because solving the proposed problems requires solving a preliminary problem first.

The problems of Book I, as said above, are elementary and like those taught in school at the time. But Diophantus treats them in his own way, using not only explicit reasoning, but also formalism; so the purpose of Book I must have been to introduce the student to a form he was not familiar with. For, as shown in Figure 6, the unknown and its (positive or negative) powers are denoted by particular symbols. There is also a symbol for "minus", actually a symbol used to separate the added terms from the subtracted terms, with the former being grouped (juxtaposed) before it and the latter after it. Accordingly, an expression like our $2x^4 - 4x^3 + 3x^2 - 1$ would then appear as $\Delta^{Y}\Delta\overline{\beta}\,\Delta^{Y}\overline{\gamma}\,\pitchfork K^{Y}\overline{\delta}\,M^{o}\overline{\alpha}$, that is, $x^4 2\ x^2 3\ -\ x^3 4\ x^0 1$ (with our x^0 transcribing the sign for units). Thus there is no sign, and no need for a sign, for the addition. Disregarding this peculiarity, both the explicit reasoning and this elementary symbolism give the translation of these problems a fairly modern appearance, at least as compared to the problems we have seen before. It is also important to note that all algebraic symbolism disappeared after Diophantus and remained absent until the fifteenth century: medieval texts, both Latin and Arabic, are purely verbal. Accordingly, the Arabic translation of Diophantus retains not the slightest trace of symbolism (which is why the reconstruction of the symbols for x^8 and x^9, which do not appear in the books preserved in Greek, is indicated as conjectural in Figure 6). Finally, note that we do not know what part Diophantus played in the introduction of this symbolism, only that, in any event, the origin of these symbols as tachygraphic abbreviations of full names is evident for almost all of them (supposing the right column of Figure 6 were written in capitals, see page 18). This is not surprising: numerous algebraic symbols at the beginning of modern times are also seen to stem from abbreviations.

units	M^{o}	μονα(δε)ς
x	ϛ	ἀριθμός
x^2	Δ^{Y}	δύναμις
x^3	K^{Y}	κύβος
x^4	$\Delta^{\mathrm{Y}}\Delta$	δυναμοδύναμις
x^5	$\Delta\mathrm{K}^{\mathrm{Y}}$	δυναμόκυβος
x^6	$\mathrm{K}^{\mathrm{Y}}\mathrm{K}$	κυβόκυβος
x^7	—	—
x^8	$\Delta\mathrm{K}\mathrm{K}^{\mathrm{Y}}$?	δυναμοκυβόκυβος ?
x^9	$\mathrm{K}^{\mathrm{Y}}\mathrm{K}\mathrm{K}$?	κυβοκυβόκυβος ?
$\frac{1}{x}$	$ϛ^{\times}$	ἀριθμοστόν
$\frac{1}{x^2}$	$\Delta^{\mathrm{Y}\times}$	δυναμοστόν
...	...	...
square	$\square^{ος}$	τετράγωνος
$=$	$ι^{σ}$	ἴσος
$-$	⋀	(λείπειν)

FIGURE 6

The *Arithmetica* is not a work of theory but a collection of problems. Primarily, these problems are indeterminate and quadratic (or reducible to a quadratic problem), with solutions that must be positive and rational (not necessarily integral, as the modern use of the expression "Diophantine equation" would imply). The few glimpses of theory contained in the *Arithmetica* appear in conditions that may be necessary for obtaining a rational solution; Diophantus mentions such conditions at the outset, immediately following the statement. Indeed, the organization of each problem is invariably as follows. First one finds the *statement*, specifying which are the given quantities (not yet set) and which are the quantities to be determined. Next comes, where necessary, the *condition* that the given quantities must satisfy for the solution to be rational. The given quantities are then *set* in accordance with the condition. The *solution* that follows begins by expressing the quantities to be determined in terms of the unknown, and then the problem is solved. One may observe the presence of these steps in the examples below.

As we have said, there is hardly any theory in the *Arithmetica*. In its introduction, however, we find a few pieces of advice on methodology. Thus, Diophantus recommends that the reader work through the problems again himself—or problems of the same kind—whereby he will grasp the particularities of the treatments. He also specifies what form the final equations should have and how to attain it:

- First, as a sort of golden rule for Diophantus's solution method, one must set the required quantities in terms of the unknown in such a way that the final equation is reduced, as far as possible, to an equality between two terms, ideally containing consecutive powers of the unknown as this guarantees the rationality of the solution. It is in this setting of the required quantities in terms of the unknown that Diophantus's art manifests itself: a suitable relation allows the problem to be transformed into a simple solution equation, whereas most other relations would not. The case in which a reduction to two terms is not possible will be considered, he says, in the later part of the *Arithmetica* (this must have been studied in the three missing books).
- Second, the final form of the equation is obtained by means of two operations. One of them is to add to both sides the (positive) amount of the subtracted terms, thus obtaining an equation involving only positive terms. The other is to remove from both sides the common quantities, whereby the equation will contain a single term for each power of the unknown. Thus, as a modern illustration, $5x^2 + 18 - 11x = 16$ would first be changed by addition of $11x$ to both sides to $5x^2 + 18 = 16 + 11x$, and then, by removal of 16, to $5x^2 + 2 = 11x$.

Remark.

- In his introduction to the *Arithmetica*, Diophantus explains that, in order to reduce the equation to its final form, we are to apply the two operations as described above. (In his words: δεήσει προσθεῖναι τὰ λείποντα εἴδη ἐν ἀμφοτέροις τοῖς μέρεσιν, ἕως ἂν ἑκατέρων τῶν μερῶν τὰ εἴδη ἐνυπάρχοντα γένηται, καὶ πάλιν ἀφελεῖν τὰ ὅμοια ἀπὸ τῶν ὁμοίων, ἕως ἂν ἑκατέρῳ τῶν μερῶν ἓν εἶδος καταλειφθῇ). Within the text, he tells us, if anything, to "add in common what is subtracted" (κοινὴ προσκείσθω ἡ λεῖψις) and "(to subtract) like from like" (ἀπὸ ὅμοιων ὅμοια). Thus he has no proper name to designate each of these two operations. However, they were later referred to by specific names in Arabic: *al-jabr* ("the restoration") denoted the first, as it "restores" the deficient side, while *al-muqābala* ("the opposition" or "the comparison") denoted the second. Since they are characteristic of algebraic manipulation, algebra came to be designated in Arabic as the "science of restoration and opposition" (*'ilm al-jabr wa'l-muqābala*). When it reached the Christian world in the twelfth century, it became in Latin, by translation, the *scientia restaurationis et oppositionis* or, by simply transcribing the Arabic name of the first of these two operations, the science of *algebra*.

Diophantus's Methods

We alluded above to the final form of equations, ideally containing only two terms involving consecutive powers of the unknown, whereby the rationality of the equation's solution is guaranteed. This can always be fulfilled in a few particular cases of the general quadratic indeterminate equation.[32]

Thus consider the indeterminate quadratic equation $ax^2 + bx + c = \square$ in general form, with a, b, c rational, in which we need to find a positive and rational value of x that will make the left side a numerical, integral or fractional, square.

(1) If $a = 0$, then the equation is reduced to $bx + c = \square$. If b and c are not both negative, set $\square = m^2$. This yields
$$x = \frac{m^2 - c}{b}.$$
Any rational m yields a rational solution x; moreover, the solution will be positive if m is chosen appropriately, according to the signs of b and c.

(2) If $c = 0$, the equation is reduced to $ax^2 + bx = \square$. If we set $\square = m^2x^2$, the equation will contain only two consecutive powers of x, and we obtain
$$x = \frac{b}{m^2 - a}.$$
Here we again have a rational solution, which will be positive with an appropriate choice of m. As in the first case, there are infinitely many choices of m yielding a positive value of x.

(3) If either a or c is a square, it is once again possible to reach a rational solution via an elementary method. It suffices to make the square term appear as part of the indeterminate square, as eliminating the square term from each side will yield an equality between two consecutive powers of x, thus a rational value for x.

Indeed, let $A^2x^2 + bx + c = \square$. Setting $\square = (Ax + m)^2$, we obtain
$$x = \frac{m^2 - c}{b - 2mA}.$$
Similarly, if $ax^2 + bx + C^2 = \square$, we set $\square = (mx + C)^2$, thus finding that
$$x = \frac{b - 2mC}{m^2 - a}.$$
In both cases, each suitable choice of m will yield a positive x.

(4) The above cases show that the solution—if there is one—is not unique, with infinitely many choices of m yielding infinitely many values of x. While Diophantus was satisfied with finding a single

[32]The linear indeterminate equation $ax + by = c$, with a, b, c integers and an integral solution required, often called today "Diophantine equation", does *not* appear in the *Arithmetica* (and would not: it has no bearing on its subject).

answer, he knew that he could obtain others as desired. He mentions this occasionally, as in Problem III.12, the twelfth problem of Book III: Diophantus asserts that finding two squares having 48 as their difference "is easy" (it is the object of II.10, see below) "and can be made in infinitely many ways" (τοῦτο δὲ ῥᾴδιον καὶ ἀπειραχῶς γίνεται). In some cases he also uses the answer obtained to find a new one, when it turns out that the one already obtained, despite being positive and rational, is not an acceptable answer for the proposed problem (for instance when it is beyond some numerical limit fixed at the outset). For, as Diophantus teaches in the lemmas to Problems "VI".12 and 15, one can obtain additional solutions as follows. Suppose that x_0 satisfies the equation

$$ax^2 + bx + c = \square,$$

so that $ax_0^2 + bx_0 + c = d^2$. If we set $x = y + x_0$, the equation becomes

$$\begin{aligned} ay^2 + (2ax_0 + b)y + (ax_0^2 + bx_0 + c) \\ = ay^2 + (2ax_0 + b)y + d^2 = \square. \end{aligned}$$

This brings us to the situation of Case (3). We can thus obtain infinitely many solutions $y(m)$, and, therefore, infinitely many solutions $x(m) = y(m) + x_0$ to the original equation.

These are some basic points that one must grasp to understand Diophantus's solution methods. And, indeed, Diophantus does not always go into all the details: as he points out in the introduction to the *Arithmetica*, the reader must rework the solutions for himself because in doing so he will understand not only the progression of the calculations but also the choice of the form taken by the quantities set or of the values of the constants introduced.

Remark.

- We have seen the reduction for $a = 0$ and $c = 0$ (first two cases above). Now suppose that $b = 0$, in which case the equation is $ax^2 + c = \square$. Whether we set $\square = m^2$ or $\square = m^2x^2$, we obtain an equality between x^2 and a constant. The rationality of the solution then depends on the coefficients a and c. Thus, in his problem "VI".14, Diophantus obtains the equation $15x^2 - 36 = \square$, which he declares impossible (καὶ αὕτη μὲν ἡ ἰσότης ἀδύνατός ἐστι διὰ τὸ τὸν ιε̅ μὴ διαιρεῖσθαι εἰς δύο τετραγώνους); indeed, as he says, a rational solution would imply that 15 is representable as a sum of two squares, which it is not. We will see the reason for this later, in our last example of a Diophantine problem. For now, what is important to observe is the connection that such indeterminate algebraic problems may have with number theory: the influence of Diophantus lies more in the fundamental questions that his problems may pose than in his treatments, which, although dexterous, are often applicable only to the given problem. In fact, this will determine our selection of problems: we will only consider those with a general solution method.

Examples of Problems

The *Arithmetica* contains about 260 problems. We will present only seven, of which four are from Book II, two use methods from Book II, and the last establishes a condition for solving one of the problems from Book II. Although this sample may seem rather unbalanced, it has the advantage of being representative and characteristic of Diophantus's algebra. The original text of three of these problems appears in Appendix B[33].

Example 1. *Arithmetica* II.8*b* (Diophantus's second solution) [Appendix B.4]
To divide a proposed square into two squares.
Let it be proposed to divide 16 *into two squares.*

Let the side of the first be x*, and the side of the other be any number of* x*'s minus as many units as the side of the (number) to be divided; let it be* $2x - 4$*. Thus, one of the squares will be* x^2*, the other* $4x^2 + 16 - 16x$*.*[34] *I also want their sum to be equal to* 16*. Thus* $5x^2 + 16 - 16x$ *is equal to* 16*, so* x *is* $\frac{16}{5}$*. The side of the first will be* $\frac{16}{5}$*, thus itself* $\frac{256}{25}$*; the side of the second will be* $\frac{12}{5}$*, thus itself* $\frac{144}{25}$*. The proof is clear.*

We thus need to solve $u^2 + v^2 = 16$. With $u = x$ and $v = mx - 4$, we obtain

$$x^2 + (mx - 4)^2 = 16,$$

and so $x = \frac{8m}{m^2+1} = u$ (and $v = \frac{4(m^2-1)}{m^2+1}$). Diophantus's solution is obtained by taking $m = 2$.

Remarks.

- We have noted that Diophantus mentions restricting conditions when necessary. As none are given here, we can deduce that any square can be represented as a sum of two rational squares and, by varying the choice of m, in infinitely many ways. The above solution makes this clear.

---

[33]For the Greek text (with Latin translation) see P. Tannery, *Diophanti Alexandrini Opera omnia cum Graecis commentariis* (2 vol.), Leipzig 1893–95 (reprint: Stuttgart 1974). The four books extant in Arabic are published in J. Sesiano, *Books IV to VII of Diophantus' "Arithmetica" in the Arabic translation attributed to Qusṭā ibn Lūqā*, New York 1982. Various studies on the Greek books are collected in Volume II of Tannery's *Mémoires scientifiques* (17 vol.), Toulouse and Paris 1912–50 (reprint: Paris 1995–). T. Heath, *Diophantus of Alexandria, a study in the history of Greek algebra*, 2nd ed., Cambridge 1910 (reprint: New York 1964) is a very complete study of the Greek books that incorporates much of Tannery's work (often without citation).

[34]As mentioned above (page 32), the subtracted terms follow the positive terms.

- As was already known in antiquity, the general integral solution of the indeterminate equation $u^2 + v^2 = w^2$ is, up to a multiplicative factor, $u = 2pq$, $v = p^2 - q^2$, $w = p^2 + q^2$, with p and q integers, and any rational multiple of such a solution is a solution as well. In our case, where $w = k$ is fixed, multiplying the above general solution by $\frac{k}{p^2+q^2}$ yields, as the general solution of the proposed problem for k given, $\frac{2kpq}{p^2+q^2}$, $\frac{k(p^2-q^2)}{p^2+q^2}$, k. If we write $\frac{p}{q} = m$ (m thus rational), this solution takes the form $\frac{2km}{m^2+1}$, $\frac{k(m^2-1)}{m^2+1}$, k. Taking $k = 4$ yields the general solution to the proposed problem, with any rational m corresponding to a solution. This shows that Diophantus's solution is general and expresses all possible rational values of the legs of a right triangle with hypotenuse length k.
- It was as a commentary to this problem that Fermat, the most illustrious disciple of Diophantus, noted that an equation of the form $x^n + y^n = z^n$ has no rational solution when n is an integer greater than or equal to 3. He asserted that he had a "truly marvelous proof" that the margin of his copy of the 1621 edition of Diophantus was too narrow to contain[35]. As a result of the (relative) narrowness of this margin, mathematicians were tormented for three centuries. In the mid-nineteenth century, E. Kummer proved Fermat's assertion for almost all exponents up to $n = 100$. In 1908, P. Wolfskehl requested in his will that the Academy of Science in Göttingen institute a prize of one hundred thousand marks for significant contributions to progress towards a solution. Via modifications of Kummer's proof, and then by use of a computer, the limit of exponents excluded was gradually extended to approximately 10^5. But it was not until the years 1993–95 that the conjecture was finally proved to be true by A. Wiles, with the help of his student R. Taylor. But what was the proof, one might ask, that Fermat himself called "truly marvelous"? Since nothing of it remains, we have no idea. However, it is likely that Fermat's proof was in fact only partial, applying only to certain exponents, and that Fermat hastily generalized his result.

Example 2. *Arithmetica* II.9 [Appendix B.5]
To divide a given number, which is the sum of two squares, into two other squares.

---

[35] *Cubum autem in duos cubos, aut quadratoquadratum in duos quadratoquadratos, & generaliter nullam in infinitum ultra quadratum potestatem in duos (sic) eiusdem nominis fas est dividere. Cuius rei demonstrationem mirabilem sane detexi. Hanc marginis exiguitas non caperet.* See (*ad loc.*) the republication of the 1621 edition of Diophantus, including Fermat's remarks, by Fermat's son, *Diophanti Alexandrini Arithmeticorum libri sex (...) cum commentariis C. G. Bacheti V. C. & observationibus D. P. de Fermat Senatoris Tolosani*, Toulouse 1670 or Volume I, p. 291 of the modern edition of Fermat's works, *Oeuvres de Fermat* by C. Henry and P. Tannery (4 vol.), Paris 1891–1912.

Let it be (proposed) to divide 13*, the sum of the squares* 4 *and* 9*, into two other squares.*

Take the sides of the aforesaid squares, 2 *(and)* 3*, and set as the sides of the desired squares* $x + 2$ *and a certain number of* x*'s minus as many units as is the side of the other (given square), say* $2x - 3$*. Their squares are, respectively,* $x^2 + 4x + 4$ *and* $4x^2 + 9 - 12x$*. It (still) remains that their sum must be* 13*. But this sum is* $5x^2 + 13 - 8x$*. This equals* 13*, and* x *is* $\frac{8}{5}$*.*

(Let us return) to the initial hypotheses. I had set the side of the first to be $x + 2$*, it will be* $\frac{18}{5}$*, and the side of the second to be* $2x - 3$*, it will be* $\frac{1}{5}$*. As for the squares, they will be* $\frac{324}{25}$ *and* $\frac{1}{25}$*, respectively. Their sum is* $\frac{325}{25}$*, making up the proposed* 13*.*

Consider, generally,

$$u^2 + v^2 = k = k_1^2 + k_2^2,$$

with k, k_1, and k_2 given.

By setting $u = x + k_1$ and $v = mx - k_2$, we obtain

$$x^2(m^2 + 1) - 2x(mk_2 - k_1) + (k_1^2 + k_2^2) = k,$$

and thus

$$x = \frac{2(mk_2 - k_1)}{m^2 + 1}.$$

Each rational m with $m > \frac{k_1}{k_2}$ yields a positive solution.

Remark.

- This problem shows that if a number can be written as a sum of two squares, then it can be written as such in infinitely many ways. Later (in "V".9), we will learn which numbers can be written as a sum of two squares.

Example 3. *Arithmetica* II.10
To find two square numbers having a given difference.

Consider $u^2 - v^2 = k$, with k given, for example (as Diophantus chooses) $k = 60$. Set $v = x$, $u = x + m$ (or, as Diophantus writes, "Set, as the side of one of them, x, and, as the other, x plus any number of units you wish as long as its square does not exceed, nor is equal to, the given difference." He then proceeds with his calculations using the choice $m = 3$.) We have, since this time x^2 drops out,

$$2mx + m^2 = k,$$

and so

$$x = \frac{k - m^2}{2m}.$$

Each (positive) rational m with $m^2 < k$ yields a solution. With $k = 60$ and $m = 3$, Diophantus finds that $x = 8 + \frac{1}{2} = v$ and $u = 11 + \frac{1}{2}$.

Remark.

- This shows that any rational number k can be expressed as a difference of two rational squares, and, in fact, in infinitely many ways.

Example 4. *Arithmetica* II.19
To find three squares such that the difference between the greatest and the middle is in a given ratio to the difference between the middle and the least.

We wish to solve $\frac{u^2-v^2}{v^2-w^2} = k$, with k given, for example (as Diophantus chooses) $k = 3$. Set $u = x + m$, $v = x + 1$, and $w = x$. Since the terms in x^2 cancel out, the equation becomes

$$2mx + m^2 - 2x - 1 = k(2x + 1),$$

and thus

$$x = \frac{m^2 - (k+1)}{2\left[(k+1) - m\right]}.$$

With $k = 3$ and choosing $m = 3$ (as $x > 0$ for $\sqrt{k+1} < m < k+1$), we obtain $x = 2 + \frac{1}{2}$, and thus $w^2 = 6 + \frac{1}{4}$, $v^2 = 12 + \frac{1}{4}$, and $u^2 = 30 + \frac{1}{4}$.

Remark.

- Setting $v = x + 1$ and $w = x$ (so $v - w = 1$), which makes the solution depend on a single parameter, is not actually a restriction, because any solution can be multiplied by an arbitrary rational factor: if u_0, v_0, w_0 is a solution, then $t \cdot u_0, t \cdot v_0, t \cdot w_0$ is a solution as well. Hence the difference between v and w can take any rational value t.

The four problems that we have examined can also be used to solve certain indeterminate systems, as some groups of problems in Books IV and V show. Once the required quantities have been suitably set in function of the unknown x, the proposed problems are reduced to a system of the form

$$\begin{cases} \alpha_1 x^{k+1} + \beta_1 x^k = \square \\ \alpha_2 x^{k+1} + \beta_2 x^k = \square', \end{cases}$$

the left sides of which contain two consecutive powers of the unknown. Then dividing the system by the even power (to keep the right side a square) turns the two left sides into linear expressions in x or $\frac{1}{x}$. We can thus restrict ourselves to considering systems of the form

$$\begin{cases} a_1 x + b_1 = \square \\ a_2 x + b_2 = \square'. \end{cases}$$

If $a_1 \cdot a_2$ (or, equivalently, $\frac{a_1}{a_2}$) is a square, then this brings us to the situation of one of the Problems II.8–10; if b_1 and b_2 are squares, then we can apply the method of Problem II.19.

(1) First, let $|a_1| = r \cdot s^2$ and $|a_2| = r \cdot t^2$; then $|a_2| = \frac{t^2}{s^2} \cdot |a_1| = h^2 \cdot |a_1|$. Multiplying the first equation by the square h^2 yields

$$\begin{cases} a_1 h^2 x + b_1 h^2 = \square_1 \\ a_2 x + b_2 = \square'. \end{cases}$$

If a_1 and a_2 have the same sign, subtract one of the two equations from the other, yielding that the difference of the squares $\square_1$, $\square'$ equals the known value $|b_1 h^2 - b_2|$. We can solve this by Problem II.10, and knowing one of the squares then yields the value of x.

If a_1 and a_2 have opposite signs, adding the two equations yields equality between the sum of the two squares and the known value $b_1 h^2 + b_2$. This can be solved by Problem II.9, provided that one representation of this value as a sum of two squares is known (or by Problem II.8 if this value is itself a square).

(2) Now, suppose that b_1 and b_2 are squares. Then the system takes the form

$$\begin{cases} a_1 x + \beta_1^2 = \square \\ a_2 x + \beta_2^2 = \square'. \end{cases}$$

Multiplying the two equations by the squares β_2^2 and β_1^2, respectively, yields

$$\begin{cases} a_1' x + \gamma^2 = \square_1 \\ a_2' x + \gamma^2 = \square_1', \end{cases}$$

where $a_1' = a_1 \beta_2^2$, $a_2' = a_2 \beta_1^2$, and $\gamma^2 = \beta_1^2 \beta_2^2$ are all known.

Since the solution x must be positive, the relative sizes of the three squares $\square_1$, $\square_1'$, and γ^2 can be determined according to the signs of a_1' and a_2'. This leads us to following three cases:

- If $a_1' > a_2' > 0$, then $\square_1 > \square_1' > \gamma^2$, and
$$\frac{\square_1 - \square_1'}{\square_1' - \gamma^2} = \frac{a_1' - a_2'}{a_2'}.$$
- If $a_1' > 0 > a_2'$, then $\square_1 > \gamma^2 > \square_1'$, and
$$\frac{\square_1 - \gamma^2}{\gamma^2 - \square_1'} = \frac{a_1'}{|a_2'|}.$$
- If $0 > a_1' > a_2'$, then $\gamma^2 > \square_1 > \square_1'$, and
$$\frac{\gamma^2 - \square_1}{\square_1 - \square_1'} = \frac{|a_1'|}{|a_2'| - |a_1'|}.$$

As each time the quantity on the right side is known, we can use Problem II.19 to determine three squares u^2, v^2, w^2 such that the quotient of the differences equals this known value. In our case, one of the squares is given (γ^2), so we then need only to multiply the answer obtained by an appropriate factor to obtain the desired solution.

Example 5. *Arithmetica* IV.42
We wish to find two numbers, one cubic and the other square, such that both the sum and the difference of the cube of the cube and the square of the square are square numbers.

The problem here is to solve

$$\begin{cases} \left(u^3\right)^3 + \left(v^2\right)^2 = \square \\ \left|\left(u^3\right)^3 - \left(v^2\right)^2\right| = \square'. \end{cases}$$

Set $u = mx$ and $v = nx^2$ with m and n chosen (Diophantus takes $m = 2$, $n = 4$). We thus obtain

$$\begin{cases} m^9x^9 + n^4x^8 = \square \equiv p^2x^8 \\ \left|m^9x^9 - n^4x^8\right| = \square' \equiv q^2x^8, \end{cases}$$

so

$$\begin{cases} m^9x + n^4 = p^2 \\ \left|m^9x - n^4\right| = q^2. \end{cases}$$

Thus $x = \frac{p^2-n^4}{m^9}$ and either $x = \frac{q^2+n^4}{m^9}$ or $x = \frac{n^4-q^2}{m^9}$.

Equating either the first two expressions for x or the first and the last leads us to finding either two squares with known difference, $p^2 - q^2 = 2n^4$ (to be solved by Problem II.10), or two squares with known sum, $p^2 + q^2 = 2n^4$ (to be solved by Problem II.9, since one representation is known).

Example 6. *Arithmetica* V.1

A fourth power, when augmented or diminished by given multiples of a cube, must give a square.

The system considered here is

$$\begin{cases} u^4 + k \cdot v^3 = \square \\ u^4 - l \cdot v^3 = \square' \end{cases} \qquad k = 4,\ l = 3.$$

Set $u = x$ and $v = qx$, for example with $q = 2$. We thus obtain

$$\begin{cases} x^4 + 32x^3 = \square \\ x^4 - 24x^3 = \square', \end{cases}$$

or

$$\begin{cases} 1 + 32r = \square_1 \\ 1 - 24r = \square_1' \end{cases} \qquad \text{with } r = \frac{1}{x}.$$

As $\square_1 > 1 > \square_1'$, then we find

$$\frac{\square_1 - 1}{1 - \square_1'} = \frac{4}{3}.$$

We can then apply Problem II.19 to determine three values $z + m$, $z + 1$, z such that

$$\frac{(z+m)^2 - (z+1)^2}{(z+1)^2 - z^2} = \frac{4}{3},$$

which yields

$$z = \frac{m^2 - \frac{7}{3}}{2\left(\frac{7}{3} - m\right)} \qquad \text{with } \sqrt{\frac{7}{3}} < m < \frac{7}{3}.$$

Taking $m = 2$ yields the values $\frac{9}{2}, \frac{7}{2}, \frac{5}{2}$. As the square of the middle value must be 1, we multiply this solution by $\frac{2}{7}$; then $\square_1 = \left(\frac{9}{7}\right)^2$, $\square_1' = \left(\frac{5}{7}\right)^2$, whence $r = \frac{1}{49}$, and finally $x = u = 49$, $v = 98$.

Example 7. *Arithmetica* "V".9

In Problem II.9, where we were asked to represent a number as the sum of two squares, one such decomposition was assumed to be already known. In addition, finding any other representation sufficed. Problem "V".9 imposes a restriction: the two squares to be found must lie within specific limits and, in fact, must be as close as possible to one another. Moreover, the problem leads to decomposing an expression depending on the given number into a sum of two squares. If this is not possible for all numbers, we cannot choose the given number arbitrarily and a condition on this given number must be specified. Following the translation of the problem (see Appendix B.6 for the Greek text), we will examine the condition and then the solution method.

To divide unity into two fractions and to add to each of the parts a given (integral) number so as to make a square.

It is then necessary that the given (number) not be odd and that the double of it increased by one unit not be divisible by a prime number which, when increased by one unit, has a fourth part.

Let it be proposed to add 6 *to each of the parts and to make a square.*

Thus, as we want to divide unity and add 6 *to each of the parts so as to make a square, then the sum of the squares is* 13. *It is thus necessary to divide* 13 *into two squares such that each of them be greater than* 6. *So if I divide* 13 *into two squares such that the difference is less than* 1, *I will solve the question. I take half of* 13, *which gives* $6 + \frac{1}{2}$, *and I look for some (square) fraction which, when added to* $6 + \frac{1}{2}$, *makes a square. Multiplying the whole by* 4, *I shall then look for a square fraction which, when added to* 26 *units, makes a square. Let the fraction to add be* $\frac{1}{x^2}$; *(then)* $26 + \frac{1}{x^2}$ *is equal to a square. Multiplying the whole by* x^2 *gives that* $26x^2 + 1$ *is equal to a square. Let its side be* $5x + 1$; *then* x *is* 10 *[as* x^2 *(is)* 100, *so* $\frac{1}{x^2}$ *(is)* $\frac{1}{100}$.][36] *Consequently, what is to be added to* 26 *will be* $\frac{1}{100}$; *thus, what is to be added to* $6 + \frac{1}{2}$ *will be* $\frac{1}{400}$, *and it makes a square, with side* $\frac{51}{20}$.

It is then necessary, in dividing 13 *into two squares, to construct the side of each as close as possible to* $\frac{51}{20}$; *and I look for that which, when diminished by* 3 *and augmented by* 2, *gives this, namely* $\frac{51}{20}$. *I then set two squares, one with (side)* $11x + 2$, *the other with (side)* $3 - 9x$. *The sum of their squares is* $202x^2 + 13 - 10x$, *equal to* 13. *Then* x *is* $\frac{5}{101}$. *Thus, the side of one of the squares will be* $\frac{257}{101}$, *and that of the other* $\frac{258}{101}$. *And if we subtract* 6 *from*

[36] A reader's (originally marginal) addition to the following sentence; it was then incorporated into the text by a later copyist.

each of their squares, it will give $\frac{5358}{10201}$ *for one of the parts of unity and* $\frac{4843}{10201}$ *for the other. It is evident that, with* 6*, each makes a square.*

The system to solve is thus

$$\begin{cases} u+v=1 \\ u+k=\square \\ v+k=\square' \end{cases}$$

for a given (integer) k.

Adding the last two equations and using the first yields

$$u+v+2k=2k+1=\square+\square'.$$

It is thus necessary for $2k+1$ to be represented as a sum of two squares. Now, if k is odd, $k=2t+1$, and we have that $2k+1=4t+3$. But no integer of the form $4t+3$ (including 3) can be represented as a sum of two squares, in integers or in fractions. If k is even, $k=2t$, and $2k+1=4t+1$. If this number is prime, it is representable as a sum of two squares; on the other hand, if it is composite, it is a sum of two squares only if any prime of the form $4m+3$ that divides it does so an even number of times. In other words, if a number N is expressed as a product of prime factors, i.e.

$$N=p_1^{\alpha_1}\cdot p_2^{\alpha_2}\cdot\ldots\cdot p_s^{\alpha_s}$$

where the p_i are distinct (and, by the usual convention, $p_1<p_2<\cdots<p_s$) with the α_i positive integers, then any p_i of the form $4m_i+3$ must appear with an even exponent α_i. This is the necessary and sufficient condition for N to be the sum of two squares.

The condition as found in the manuscripts preserving the *Arithmetica* is unclear. It was reestablished by Fermat. The first part excludes from consideration all integers of the form $4t+3$, while the second states that the number to be decomposed must not be divisible by a prime of the form $4m_i+3$, or, according to the text, by a prime that, when increased by 1, becomes divisible by 4. It is implicitly understood that the largest square factor of the number under consideration has already been removed—as such a factor plays no role in decomposing the number into a sum of squares—and thus that the quantity left, of the form $p_1\cdot p_2\cdot\ldots\cdot p_r$, must not contain any prime of the form $4m_i+3$. The reason that Fermat had to reestablish this condition was that the text transmitted through all surviving manuscripts is corrupted at this place due to the negligence or carelessness of a copyist; however, the remaining fragments show that Diophantus did indeed know the elements of the condition as reconstructed by Fermat[37].

[37]It may seem surprising that a corruption of the text would be common to all the manuscripts. However, ancient texts that were less commonly used, as were mathematical texts of higher level, often survived at the end of antiquity through a single specimen that served as the source of all subsequent copies. See G. Toomer, "Lost Greek mathematical works in Arabic translation", *Mathematical Intelligencer*, 6 (1984), pp. 32–38; reprinted pp. 275–84 in Christianidis' *Classics* (note 6).

Taking $k = 6$ as the given number, $2k + 1 = 13 = 2^2 + 3^2$ satisfies the condition. On the other hand, since u and v must be positive and also less than 1, we must have $0 < u,\ v < 1$. Adding 6 then implies that the squares appearing on the right side must satisfy the condition $6 < \square,\ \square' < 7$. Therefore, one cannot consider just any decomposition of 13. Diophantus calls his procedure of determining two squares close to one another the "method of approaching" (παρισότητος ἀγωγή); indeed, this method attains its aim indirectly, via the intermediate search for a single square falling within the same bounds as the two desired squares.

As 13 is the sum of two squares between 6 and 7, we look for a square near its half, $6 + \frac{1}{2}$:

$$6 + \frac{1}{2} + \text{small square fraction}^{38} = \text{square}.$$

Multiplying both sides by the square factor 4 and writing the additive fraction as $\frac{1}{y^2}$ yields[39]

$$26 + \frac{1}{y^2} = \text{square},$$

and so

$$26y^2 + 1 = \text{square}.$$

As Diophantus does, we set the square to be $(5y + 1)^2$.[40] Then $y = 10$, thus $26 + \frac{1}{100}$ is a square, as is its quarter $6 + \frac{1}{2} + \frac{1}{400}$, which equals $\left(\frac{51}{20}\right)^2$.

Now we can turn to the stated problem, where the goal is to determine two squares close to $\left(\frac{51}{20}\right)^2$ that add up to $13 = 2^2 + 3^2$. To achieve this, Diophantus begins by expressing $\frac{51}{20}$ in terms of the values 2 and 3:

$$\frac{51}{20} = 3 - 9 \cdot \frac{1}{20}$$
$$\frac{51}{20} = 2 + 11 \cdot \frac{1}{20},$$

and then replaces the common fractional term $\frac{1}{20}$ by x, which gives the final equation

$$(3 - 9x)^2 + (2 + 11x)^2 = 13.$$

This causes the constant terms to drop out, and thus, as desired, we are left with a linear equation: $202x = 10$, whence $x = \frac{5}{101}$. Diophantus thus finds

[38]The word "square" does not appear in the Greek text; perhaps it was missed by a copyist.

[39]Here we write y where Diophantus, who has only a single symbol for the unknown, writes the equivalent of our x. He thus uses the same unknown for the main problem (see below) and this preliminary problem. But no confusion between these unknowns is possible as the two calculations are clearly separate.

[40]Diophantus does not justify his choice. However, if we set the square to be $(my + 1)^2$, we obtain $y = \frac{2m}{26-m^2}$; then $m = 5$ is indeed the most appropriate integer value, because y must be large for the fraction added to 26 to be small. Remember that Diophantus intended the student to rework the problems in order to thoroughly understand the solutions (see pages 33–34, 36).

that $\square = (3-9x)^2 = \left(\frac{258}{101}\right)^2$, $\square' = (2+11x)^2 = \left(\frac{257}{101}\right)^2$, and $u = \frac{5358}{10201}$, $v = \frac{4843}{10201}$.

Remark.

- The same method can be used to determine three squares close to one another that add up to a given quantity. First, we find a square near $\frac{1}{3}$ of the given quantity and then set the final equation as above, by introducing the terms of the known decomposition as constants in the three terms of the trinomial. This problem is solved by Diophantus after the one above (Problem "V".11). In this case as well, he specifies the form that the quantity to be partitioned must not take (positive integers of the form $8t+7$ cannot be written as a sum of less than four squares, either in integers or in fractional numbers).

The Later Influence of Greek Indeterminate Algebra

Some problems from Diophantus's *Arithmetica* can be found in Arabic texts from as early as the tenth century, and a few authors clearly assimilated his methods. Thus, the Persian mathematician Karajī (around 1010), while he simply reproduces a sequence of problems from the first four of Diophantus's books in his *Fakhrī*, presents in his *Badī'* a synthesis of Diophantus's general methods[41]. On the other hand, in the main treatise by the Egyptian Abū Kāmil (around 880), the *Algebra* (Arabic title: *kitāb fi'l-jabr wa'l-muqābala*), one finds a section containing methods that we would call Diophantine; however, although certainly Greek in origin, the problems presented are different in form and sometimes in substance from those found in the *Arithmetica* (which, in fact, had not yet been translated or had only just been translated at the time). In this Abū Kāmil appears to be carrying on a legacy of Ancient Alexandria, as some of his colleagues must have already done before him since he informs us that such problems were the object of discussion among mathematicians of his time[42].

Among the equations or indeterminate systems that Abū Kāmil solves in his *Algebra*, the four cases below are not known from the extant *Arithmetica*—and this may be what Diophantus treated in the three lost books, of which we only know from his introduction that they considered situations where the reduction of the solution equation to an equality between two

[41]F. Woepcke, *Extrait du Fakrî*, Paris 1853 (reprinted by the Frankfurt Institut für Geschichte der arabisch-islamischen Wissenschaften in 1986 in F. Woepcke, *Etudes sur les mathématiques arabo-islamiques*, 2 vol.). A. Anbouba, *L'Algèbre al-Badī' d'al-Karagī*, Beirut 1964. For a study of the indeterminate algebra in the latter work, see J. Sesiano, "Le Traitement des équations indéterminées dans le *Badī' fī'l-Ḥisāb* d'Abū Bakr al-Karajī," *Archive for history of exact sciences*, 17 (1977), pp. 297–379.

[42]See J. Sesiano, "Les Méthodes d'analyse indéterminée chez Abū Kāmil," *Centaurus*, 21 (1977), pp. 89–105.

terms was not immediately possible (page 34). This is exactly the case that some of the problems below address.

(1) Consider the equation

$$-x^2 + 2bx + c = \square$$

with b and c given and at least one of them positive. This is solvable, Abū Kāmil says, if $b^2 + c$ is representable as a sum of two squares (if $c < 0$ we must have that $b^2 > |c|$; otherwise, "the problem is impossible"). Indeed, $x = b + \sqrt{b^2 + c - \square}$, so if $b^2 + c = p^2 + q^2$, then setting $\square = p^2$ yields the rational solution $x = b + q$ and thus, as we have seen (page 36), as many others as desired.

(2) The indeterminate quadratic system

$$\begin{cases} x^2 + bx + c = \square \\ x^2 + bx + c + h\sqrt{x^2 + bx + c} = \square', \end{cases}$$

with b, c, and h given, can be solved by setting $\square' = (x+m)^2$, with m to be determined. Then

$$x^2 + bx + c + h\sqrt{x^2 + bx + c} = x^2 + 2mx + m^2,$$

and so

$$h\sqrt{x^2 + bx + c} = x\,(2m - b) + \left(m^2 - c\right).$$

Squaring each side then yields

$$h^2x^2 + h^2bx + h^2c$$
$$= x^2\,(2m - b)^2 + 2x\,(2m - b)\left(m^2 - c\right) + \left(m^2 - c\right)^2.$$

We will obtain a rational x satisfying both equations by taking

$$m = \frac{b \pm h}{2}.$$

This eliminates the x^2 terms, leaving

$$x = \frac{\left(m^2 - c\right)^2 - h^2c}{h^2b - 2\,(2m - b)\,(m^2 - c)}.$$

Here, however, the sign of x will depend on the given values.

(3) Abū Kāmil also treats the system

$$\begin{cases} x^2 + bx = \square \\ x^2 + bx + h\sqrt{x^2 + bx} = \square'. \end{cases}$$

For this system to have a solution, he says, we need h and b to have the same sign, with $h > b$ if $h, b > 0$ and $|h| < |b|$ if $h, b < 0$.

Considering the first equality, he sets $\square = \left(\frac{h}{b}\right)^2 x^2$ to obtain

$$x_0 = \frac{b}{\left(\frac{h}{b}\right)^2 - 1} = \frac{b^3}{h^2 - b^2}.$$

This positive value x_0 is a solution of the system, because it satisfies the second equation as well:

$$\begin{aligned} x^2 + bx + h\sqrt{x^2+bx} &= \square + h\sqrt{\square} \\ &= \left(\frac{h}{b}\right)^2 x_0^2 + h\left(\frac{h}{b}\right) x_0 \\ &= \left(\frac{h}{b}\right)^2 x_0^2 + b\left(\frac{h}{b}\right)^2 x_0 \\ &= \left(\frac{h}{b}\right)^4 x_0^2. \end{aligned}$$

Abū Kāmil then observes that one can construct a sequence of equations that are all satisfied by the value found:

$$\begin{cases} x^2 + bx = \square_1 \\ x^2 + bx + b\left(\frac{h}{b}\right)\sqrt{x^2+bx} = \square_2 \\ x^2 + bx + b\left(\frac{h}{b}\right)\sqrt{x^2+bx} \\ \qquad\qquad +b\left(\frac{h}{b}\right)^2\sqrt{x^2+bx+b\left(\frac{h}{b}\right)\sqrt{x^2+bx}} = \square_3 \\ \dots \end{cases}$$

(4) Finally, Abū Kāmil solves the system

$$\begin{cases} x^2 + b_1 x + c_1 = \square \\ x^2 + b_2 x + c_2 = \square'. \end{cases}$$

Consider that the difference between the roots of these two squares equals a value m (to be determined):

$$\sqrt{x^2+b_1x+c_1} - \sqrt{x^2+b_2x+c_2} = m.$$

Let us eliminate the radicals. First,

$$\begin{aligned} &x^2 + b_1x + c_1 \\ &\quad = x^2 + b_2x + c_2 + 2m\sqrt{x^2+b_2x+c_2} + m^2, \end{aligned}$$

whence

$$(b_1 - b_2)\,x + \left(c_1 - c_2 - m^2\right) = 2m\sqrt{x^2+b_2x+c_2}.$$

Then,

$$\begin{aligned} &(b_1-b_2)^2\,x^2 + 2x\,(b_1-b_2)\,(c_1 - c_2 - m^2) \\ &\qquad + \left(c_1 - c_2 - m^2\right)^2 = 4m^2x^2 + 4m^2b_2x + 4m^2c_2. \end{aligned}$$

To make x rational, we set

$$m = \frac{|b_1 - b_2|}{2},$$

thus obtaining

$$x = \frac{(c_1 - c_2 - m^2)^2 - 4m^2c_2}{4m^2b_2 - 2(b_1 - b_2)(c_1 - c_2 - m^2)}.$$

Here again, however, the sign of x will depend on the given values.

We have seen that one of the most attentive readers of the *Arithmetica* was Fermat, at a time when Diophantus's reputation was again at its height. Indeed, Diophantine indeterminate algebra was then completely new to Europe, as the Middle Ages knew nothing of Diophantus's work. A few manuscripts of the six Greek books of the *Arithmetica* certainly reached Italy after the fall of Constantinople (1453), and the first mention of Diophantus appears around 1464 in a letter written by the German astronomer Johann Müller from Königsberg (Latin *Regius Mons*, whence the name Regiomontanus, by which he is better known). It took a century to see the first serious study on Diophantus, when an Italian adaptation of the first five books was made by A. Pazzi and R. Bombelli; it was not published as such, but almost all of these problems were incorporated into Bombelli's *Algebra*, which was printed in Bologna in 1572 (see Chapter 5). The first real translation of the *Arithmetica*, which appeared in 1575, is by G. Xylander, or, rather, W. Holtzmann.[43] In the words of Xylander, in it Diophantus's text is "rendered into Latin by incredible efforts, and explained by commentaries" (*incredibili labore Latinè redditum, et Commentariis explanatum*). It appears that the translation rapidly attracted attention. In his *Arithmetique*, which was written in 1582 and appeared in 1586–87, E. E. Leon Mellema writes that the *Arithmetica* is found "every day in the hands of mathematicians" (*journellement (...) entre les mains des Geometriciens & Arithmeticiens*). This is confirmed by contemporary studies, notably the one in 1585 by the Flemish mathematician S. Stevin, who is also known as one of the major proponents of the use of decimal fractions. An edition of the Greek text had not yet been published; this gap was filled in 1621, with an edition in which Diophantus's original text is accompanied by a Latin translation and careful commentary by C. G. Bachet de Méziriac. It was in the margin of a copy of this edition that Fermat wrote, or was unable to write in full, his remarks and additions.

Yet, after Fermat and his contemporaries, mathematical interest in Diophantus and problems of this genre waned. It disappeared altogether with the coming of new mathematical methods and an emphasis on studies concerning conditions for the existence of a solution rather than numerical answers (Gauss's *Disquisitiones arithmeticae*, published in 1801, marks this rupture). One of the last mathematicians still concerned with Diophantus's methods is the Swiss L. Euler (1707–1783), particularly in the second part of his *Vollständige Anleitung zur Algebra* ("Complete instruction in algebra",

[43]Antiquity was admired so greatly during the Renaissance that several editors and translators of Greek works adopted Hellenic versions of their names.

or, as the English edition calls it, *Elements of Algebra*).[44] There, Euler treats indeterminate algebra and the book contains, in addition to generalizations of Diophantus's methods, solutions for the following two cases of the indeterminate quadratic equation $ax^2 + bx + c = \square$. (The two cases a or c a square, already treated by Diophantus, are considered by Euler just before the two below.)

Case (1) $ax^2 + bx + c = \square$, where the discriminant $b^2 - 4ac$ is a square.

If $b^2 - 4ac = d^2$, then the roots of the equation $ax^2 + bx + c = 0$ are

$$x_{1,2} = \frac{-b \pm d}{2a},$$

and we have that

$$ax^2 + bx + c = a\left(x + \frac{b-d}{2a}\right)\left(x + \frac{b+d}{2a}\right).$$

We need to make this latter expression a square. Consider, generally,

$$ax^2 + bx + c = (f + gx)(h + kx).$$

By setting

$$(f + gx)(h + kx) = \frac{m^2}{n^2}(f + gx)^2,$$

we obtain

$$x = \frac{fm^2 - hn^2}{kn^2 - gm^2}.$$

Euler followed Diophantus in choosing m and n so as to obtain a positive solution.

Example 1. *Vollständige Anleitung*, II, 2, 53
Consider the equation

$$6x^2 + 13x + 6 = \square.$$

As $d^2 = 25$, we have

$$\begin{aligned} 6x^2 + 13x + 6 &= (2 + 3x)(3 + 2x) \\ &= \frac{m^2}{n^2}(2 + 3x)^2, \end{aligned}$$

[44]Euler's *Algebra* has an unusual history. It was composed while Euler, already blind in one eye, was gradually losing sight in his remaining eye. He thus decided to dictate to his valet a work on algebra that would be broadly accessible and that would teach algebra from first principles. This valet was certainly representative of the general public: he had no particular training in mathematics and was of average intelligence (*gehörete*, writes the editor, *was seine Fähigkeit anlanget, unter die mittelmäßigen Köpfe*). Over time, however, the valet acquired a good understanding of contemporary algebra; he doubtlessly owed this to his perseverance, but also to the extraordinary clarity of Euler's exposition, present even in a work composed from memory.

so

$$x = \frac{3n^2 - 2m^2}{3m^2 - 2n^2}.$$

For $x > 0$ we must take $\frac{2}{3} < \frac{m^2}{n^2} < \frac{3}{2}$. We can then choose $m = 6$, $n = 5$ (as $\frac{36}{25}$ indeed falls within the prescribed limits), yielding $x = \frac{3}{58}$.

Case (2) $ax^2 + bx + c = \square$, where this time the discriminant $b^2 - 4ac$ is not a square, but where $ax^2 + bx + c$ can be written in the form $(f_1 + g_1x)^2 + (f_2 + g_2x)(f_3 + g_3x) \equiv p^2 + qr$.

In this case, we write

$$p^2 + qr = \left(p + \frac{m}{n}q\right)^2,$$

so that

$$p^2 + qr = p^2 + 2\frac{m}{n}pq + \frac{m^2}{n^2}q^2,$$

and thus

$$n^2r = 2mnp + m^2q.$$

This linear equation yields the value of x.

Example 2. *Vollständige Anleitung*, II, 2, 55 [Appendix E.1]
To solve

$$2x^2 - 1 = \square,$$

set

$$x^2 + (x+1)(x-1) = \left[x + \frac{m}{n}(x+1)\right]^2;$$

then

$$x = \frac{m^2 + n^2}{n^2 - 2mn - m^2}.$$

For m, n, Euler successively chooses the pairs -1, 1; -1, 2; -1, -2. This last choice separates him significantly from the Ancients, who would not have considered a negative x (page 24); yet, as Euler says, "since in our equation $2x^2 - 1$ only the square x^2 occurs, then it does not matter whether the values found for x are positive or negative."

Chapter 3

Algebra in the Islamic World

3.1. Introduction

When the ʿAbbāsid dynasty replaced that of the Umayyads, Baghdad (which was founded in 762) came to replace Damascus as the political center of the Islamic world. It became its cultural center as well, with the creation of the House of Wisdom (*bayt al-ḥikma*), a sort of academy that carried on the tradition of the major intellectual centers of antiquity, Athens and Alexandria. Its early scientific activities consisted primarily in unifying, over the course of a century, a science resulting from three very different scientific heritages.

The first of these was the heritage of Mesopotamia. A direct transmission, such as the Greeks had benefited from, was no longer possible: as the use of cuneiform writing had died out in the second century, the understanding of tablets using this writing was lost until modern times. Nonetheless, Mesopotamian knowledge survived partially through its integration into Greek knowledge, and it may sometimes even have been transmitted further as such: the Persian scholar Bīrūnī (972–1048) mentions that a certain procedure for measuring the length of daylight came from the "Babylonians" (*ahl Bābil*)[45].

This question of a masked transmission does not arise for the heritage of India, as not only was Indian science alive and even still at the dawn of its evolution, but we in fact have proof of its early encounter with the Muslim world, through an official invitation of Indian scholars to Baghdad soon after the founding of the House of Wisdom. This Indian heritage, transmitted at about the same time, is particularly evident in two domains. First, it marked early astronomical treatises, where it is found intermixed with Greek elements (which were either already used in India or incorporated by the Muslims). In particular, what survived of this Indian influence, even after the reception of classical Greek astronomy, was trigonometry. In the beginning,

[45] Al-Bīrūnī, *Ifrād al-maqāl fī amr al-ẓilāl*, Hyderabad 1948, p. 138; E. Kennedy, *The exhaustive treatise on shadows by (...) al-Bīrūnī* (2 vol.), Alep 1976, I (translation), p. 186, II (commentary), pp. 114–115.

it was considered to be a science auxiliary to astronomy, but it gradually earned its autonomy with Muslim science[46]. The second domain depending on the influence of Indian mathematics was the system of writing numbers using 10 particular symbols, which we call "Arabic numerals," but which the Arabs called "Indian numerals" (*ḥurūf al-hind*), with the recipients in both cases acknowledging the donors. In addition to the symbols of the digits, their use in arithmetic was adopted ("Indian arithmetic", *ḥisāb al-hind*). It would seem that the extensive application of mathematics to practical life, and particularly to commerce, is of Indian origin as well. Moreover, the acquisition of this utilitarian mathematics was facilitated by the simplicity of the arithmetical operations in the ten-digit system, making it accessible to a broader public. Indeed, it was due to the needs of commercial mathematics that arithmetic and algebra spread so rapidly during the Middle Ages.

These two heritages alone would not have been enough to yield progress in the theoretical exact sciences. From Greece, the Islamic scientists received a way of thinking in terms of definitions, theorems, and proofs, both in geometry (Euclid, Archimedes; Apollonius for conic sections) and in applied sciences such as spherical trigonometry (Menelaus), mechanics (Archimedes, Heron), optics (Euclid, Ptolemy), and astronomy (Ptolemy). These contributions are essential: Islamic mathematicians demonstrate Greek rigor in establishing theorems through proofs, using empiricism only as a method of guessing or roughly evaluating new assertions. They were the first beneficiaries of this unique transformation of science from a synthesis of experiments to a system of logical deductions. The influence of Greek mathematics is also evident in the study of the properties of whole numbers, available in Arabic early on, namely through the translation of the thirteen books of Euclid's *Elements* (of which Books VII to IX are fundamentals of number theory) and also Nicomachus's *Introduction to Arithmetic*, a largely accessible number theory text written around A.D. 100. Finally, Arab algebra received the stamp of antiquity in two very different ways: first, from geometry, which, by means of figures, aided both the understanding of algebraic reasoning and the proving of formulas or identities; and then in indeterminate algebra, through the translation of Diophantus or the use of other unknown Alexandrian sources (see above, page 46).

But Greek science, at least at a higher level, survived only in manuscripts. The Islamic world encountered it predominantly during the ninth century. All that could be translated was translated, from minor compilations of late antiquity to the great classical treatises. The latter were the object of particular care, and today we know of scholars and translators who, when faced with an incomplete or poor-quality manuscript, endeavored to find

[46]Sine and cosine were used in India; the Muslims introduced the other trigonometric functions. Plane trigonometry had a different form in antiquity: the Greeks used a system based on the relation between an arc and its whole chord.

better copies[47]. The early translations were then improved and commented upon, and it can be said that Greek science—or, rather, what had been transmitted of it—was fully assimilated by the middle of the tenth century. We will follow one aspect of this growing influence of Greek geometry on algebra through two authors of algebraic works: one, al-Khwārizmī, made use of an intuitive sort of geometry in the early ninth century, while the other, Abū Kāmil, applied Euclid's theorems at the end of the same century.

Although these two authors differ in their knowledge, or at least in their use, of Euclid's work, they have in common their algebraic language, which remained practically unchanged in the work of later Muslim algebraists. However, it would perhaps be more appropriate to say that what they all shared is a lack of algebraic language. Indeed, they used no symbols, instead expressing everything, from reasoning to algebraic operations, in words. Numbers were expressed in words as well: Indian digits were used only in texts on arithmetic and in tables (although the latter often used a system derived from the Greek system instead[48]). At least the unknown and its powers have specific names: the unknown x is "thing" (*shay'*), x^2 is "wealth" or "property" (*māl*, which we will express for convenience by "square" in the translations), x^3 is "cube" (*ka'b*). Higher powers are expressed using combinations of the latter two terms: if the exponent is divisible by 3, thus of the form $3m$, the power of the unknown is denoted by *ka'b* repeated m times; if it is of the form $3m+2$, one *māl* precedes *ka'b* repeated m times; if it is of the form $3m+1$, *māl* repeated twice is followed by *ka'b* repeated $m-1$ times. The successive powers from x^4 onwards are then *māl māl*, *māl ka'b*, *ka'b ka'b*, *māl māl ka'b*, *māl ka'b ka'b* (exceptionally: *māl māl māl māl*), and so forth. This clearly follows the same principle as used in Greece, which we have encountered with Diophantus. Finally, the positive terms of an equation (connected by *wa*, "and") are grouped before the negative terms (announced by *illā*, "except", "less"). This also was the case in Greece; but there is no point in speculating about some influence since separating the terms according to the signs is dictated by verbal algebra, all the terms placed after the word "minus" being subtractive. The operations for simplifying an equation, which characterized algebra before naming it, have been discussed above (page 34).

Thus the Arabs neither adopted nor even adapted the Greek algebraic symbolism of Diophantus upon learning of it. Neither did they do so with the other main system in use at the time, that of the Indians. This system likewise shows how algebraic symbolism naturally arises from an abbreviation of specific, systematically repeated words. For example, the designation of the constant term, *rūpa*, was abbreviated to *rū* in equations; the unknown, *yāvat(-tāvat)* "as much as" became *yā*; and its subsequent two powers were

[47]See G. Toomer, *Apollonius: "Conics", Books V to VII* (2 vol.), New York 1990, pp. xviii and 620–29.

[48]The units, the tens, the hundreds and one thousand were represented by the twenty-eight Arabic letters.

similarly abbreviated, with *varga* being written as *va* and *ghana* as *gha*. The higher powers were again composed from the second and third powers, but this time hardly in any consistent way: if an exponent could be decomposed as $n = 2^r 3^s$, *va* was repeated r times and *gha* was repeated s times; if not, n was decomposed as $2r + 3s$, and *va* was repeated r times and *gha* repeated s times, with the term *ghāta* being added at the end to indicate that the decomposition used was additive. Thus, the twelfth power of an unknown was *varga varga ghana*, while the seventh power was *varga varga ghana ghāta*. Different unknowns in linear problems were either numbered or distinguished by words of the same kind, such as names of colors; these too were abbreviated. Note finally that the names of the common operations also came to be abbreviated.

Remark.

- When algebraic symbolism made its first appearance in Arabic writings in the fifteenth century (just at the same time as in Europe), the symbols adopted were, here too and not unexpectedly, the initial letters of the corresponding mathematical terms.

3.2. Al-Khwārizmī

Al-Khwārizmī's *Algebra*

Muḥammad ibn Mūsā al-Khwārizmī is said in Arabic sources to be the first mathematician to write in Arabic, even the very first algebraist, and his work was of considerable influence in the medieval world. His name suggests that he was of Persian origin, but we know that he lived in Baghdad during the Caliphate of al-Ma'mūn (813–833), and it was there that he composed his works on arithmetic, algebra, astronomy, geography, the calendar, and even history. Yet his reputation is, as far as originality is concerned, largely overrated: he does not appear to have been a discoverer. However, he was an able scientist, simply but clearly presenting the material he considered, and his importance lies in the fact that he initiated and helped spread scientific study in Arabic, thus also giving rise to numerous vocations. This influence was not limited to the Arabic-speaking world: his educational impact re-emerged in medieval Europe thanks to the Latin translation of his *Arithmetic* and *Algebra* in the twelfth century. Today, we continue to honor him daily, often without even realizing it: latinized in the twelfth century from a misreading of the Arabic, his name became *algorizmus*; then *algoritmus* through a misreading of the Latin; and finally *algorithmus* through a mistakenly assumed Greek etymology. The origin of the name had soon fallen into oblivion and only its link to computational procedures was remembered.

His *Compendious Book on Algebra*—as that is what his influential *Algebra*, written around 820, appears to have been called—is, according to its author, "a short work on algebraic calculation, confining it to what is easiest and most useful in arithmetic, such as men constantly require in cases of inheritance, legacies, partition, lawsuits, and trade, and in all their dealings with one another, or where the measuring of lands, the digging of canals, geometrical computation, and other objects of various sorts and kinds are concerned." More precisely, the book defines the three fundamental algebraic quantities (number, root, square; see below), then explains the solution of linear and quadratic equations, the fundamentals of algebraic calculation, and next solves around forty problems, thus showing how to reduce a given problem to one of six standard equations. The goal of the subsequent sections is then to apply the rule of three to commerce or to the measurement of plane and solid geometric figures; and, finally, to apply algebra to the partition of inheritances[49].

The Six Equations

What al-Khwārizmī calls the "six equations" are the various forms taken by linear and quadratic equations when the terms on both sides of the equality are positive and thus a positive solution is possible. These are

- $ax^2 = bx$ ("Squares are equal to roots")
- $ax^2 = c$ ("Squares are equal to a number")
- $bx = c$ ("Roots are equal to a number")
- $ax^2 + bx = c$ ("Squares and roots are equal to a number")
- $ax^2 = bx + c$ ("Roots and a number are equal to squares")
- $ax^2 + c = bx$ ("Squares and a number are equal to roots")

None of this is new, of course, and neither is the subsequent explanation of how to solve these equations. What is distinctive, though, and what remained part of the explanations of these solutions until the beginning of modern times, is the use of a geometrical representation of the solution formulas. We will now show this for the three cases of a trinomial quadratic equation, summarizing the author's purely verbal text and using letters for the coefficients (each type is originally presented for a numerical example where the coefficient of x^2 is 1).

Case 1: $x^2 + px = q$ (Figure 7)

Let AB be the square x^2 (following the Greek custom, rectangles are often referred to, in abbreviated notation, by the two letters denoting opposite angles). Extend the sides from each endpoint by the known value $\frac{p}{4}$. The area of the larger square CD is thus equal to $\left(x + \frac{p}{2}\right)^2$. On the other hand, it is composed of nine elements with the following areas: x^2, four rectangles $\frac{p}{4} \cdot x$, and four squares $\left(\frac{p}{4}\right)^2$. Because $x^2 + 4 \cdot \frac{p}{4}x = q$, as given, and $4\left(\frac{p}{4}\right)^2 = \left(\frac{p}{2}\right)^2$, the sum

[49]Edition, with English translation, by F. Rosen, *The Algebra of Mohammed ben Musa*, London 1831 (reprint: Hildesheim 1986). There is also a (partial) Latin translation, published along with other medieval mathematical texts in the appendix to Vol. I (pp. 253 *seqq.*) in G. Libri, *Histoire des sciences mathématiques en Italie* (4 vol.), Paris 1838–41 (reprint: Hildesheim 1967).

of these nine areas is $\left(\frac{p}{2}\right)^2 + q$. Thus we have

$$\left(x + \frac{p}{2}\right)^2 = \left(\frac{p}{2}\right)^2 + q,$$

and so

$$x = \sqrt{\left(\frac{p}{2}\right)^2 + q} - \frac{p}{2}.$$

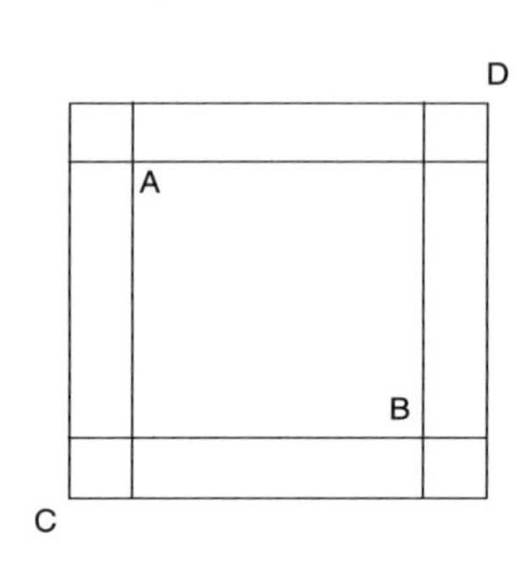

FIGURE 7

Case 2: $x^2 = px + q$ (Figure 8)

Let AD $= x^2$ and BE $= p$ $(< x)$; then AE $= px$, implying that CE $= q$. Now set BF $=$ FE $= \frac{p}{2}$, then draw FH $=$ FE perpendicular to BD, yielding HE $= \left(\frac{p}{2}\right)^2$. Next, extend FH by HK $=$ ED. As FD $= x - \frac{p}{2}$, KD $= \left(x - \frac{p}{2}\right)^2$. On the other hand, KI $=$ LG, as their corresponding sides are equal. Hence KI $+$ EG $= q$, and so KD $= \left(\frac{p}{2}\right)^2 + q$. We deduce that

$$\left(x - \frac{p}{2}\right)^2 = \left(\frac{p}{2}\right)^2 + q,$$

that is,

$$x = \frac{p}{2} + \sqrt{\left(\frac{p}{2}\right)^2 + q}.$$

Case 3: $x^2 + q = px$ (Figure 9)

Let AD $= x^2$ and BE $= p$ $(> x)$; then AE $= px$, implying that CE $= q$. Now set BF $=$ FE $= \frac{p}{2}$, then draw FH $=$ FE perpendicular to BE, yielding HE $= \left(\frac{p}{2}\right)^2$. Now set KI $=$ KH $=$ DF $= \frac{p}{2} - x$; we thus have that HI $= \left(\frac{p}{2} - x\right)^2$. On the other hand, LI $=$ KD, as their corresponding sides are equal. Hence LI $+$ KE $= q$, and so HI $= \left(\frac{p}{2}\right)^2 - q$. We deduce that

$$\left(\frac{p}{2} - x\right)^2 = \left(\frac{p}{2}\right)^2 - q,$$

that is,

$$x = \frac{p}{2} - \sqrt{\left(\frac{p}{2}\right)^2 - q}.$$

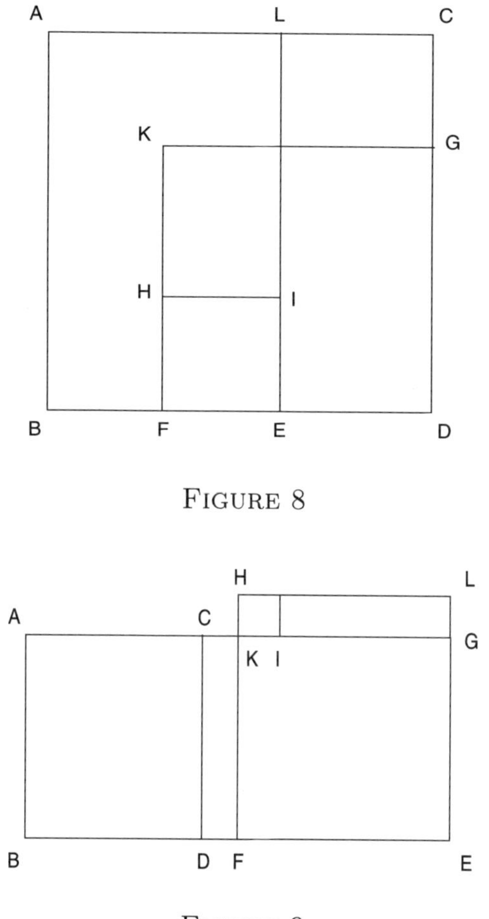

FIGURE 8

FIGURE 9

Here al-Khwārizmī implicitly assumes that BF > BD, meaning that $\frac{p}{2} > x$. However, another positive solution is possible in the third case, which al-Khwārizmī alludes to by writing that the square root could be added instead. This second solution is studied in more detail by a contemporary of al-Khwārizmī, 'Abd al-Ḥamīd ibn Turk.[50] Let AD $= x^2$ and BE $= p$; then AE $= px$, implying that CE $= q$ (see Figure 10). Now set BF $=$ FE $= \frac{p}{2}$, then draw FH $=$ FE perpendicular to BE, yielding HE $= \left(\frac{p}{2}\right)^2$. Now set FK $=$ KI $=$ FD $= x - \frac{p}{2}$; then FI $= \left(x - \frac{p}{2}\right)^2$. On the other hand, HI $=$ CL, as their corresponding sides are equal. Hence HI+DL $= q$, and thus FI $= \left(\frac{p}{2}\right)^2 - q$. We deduce that

$$\left(x - \frac{p}{2}\right)^2 = \left(\frac{p}{2}\right)^2 - q,$$

[50]Publication and English translation of his work by A. Sayılı, "Ibn Türk'ün Cebri (the Algebra of Ibn Turk)", *Türk tarih kurumu yayınlarından*, ser. 7, 41 (1962).

that is,

$$x = \frac{p}{2} + \sqrt{\left(\frac{p}{2}\right)^2 - q}.$$

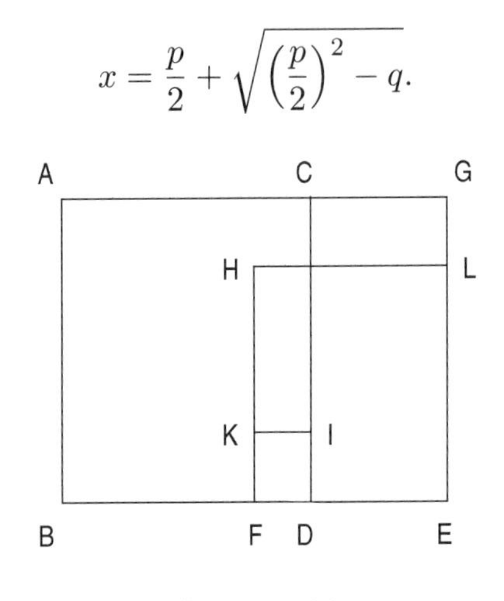

FIGURE 10

Therefore, there are two positive roots in the third case, provided that $\left(\frac{p}{2}\right)^2 - q > 0$. For the case $\left(\frac{p}{2}\right)^2 = q$, ʿAbd al-Ḥamīd includes an illustration in which the pairs of points D, F and C, H respectively coincide (Figure 11).

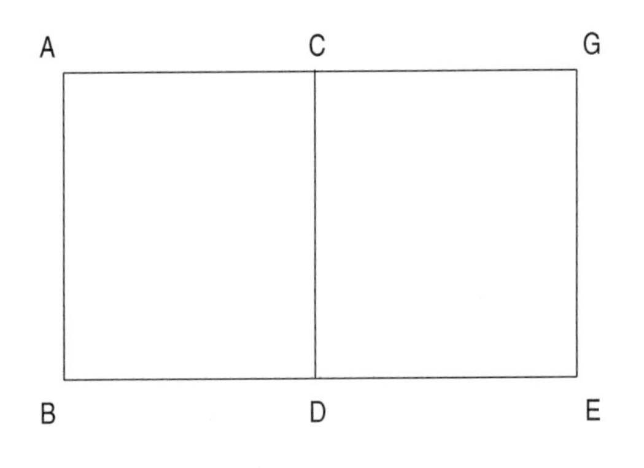

FIGURE 11

ʿAbd al-Ḥamīd's study appears to be more complete than that of al-Khwārizmī, at least in the examination of the six equations (for we do not know if this is the case for the rest of his *Algebra*, as only the fragment on the solutions of the six equations remains today). We also do not know when he wrote his *Algebra*, but this does confirm that al-Khwārizmī was not the sole keeper of a somewhat forgotten algebraic knowledge, or at least that he soon met with competition in this role. As often happens in science—just as in any other domain of knowledge—there are some particularly auspicious times when all necessary conditions come together to drive the birth or the rebirth of a theory or a discovery. Given that each person who benefits from this conjunction is aware of his own idea but not that the favorable combination may benefit others as well, it is hardly surprising that unwarranted disputes over priority and accusations of plagiarism often arise. Then, whether by conviction or by inclination, or simply wishing to flatter, subsequent generations join in, always more vehemently as events recede into the ever-darkening distance. It thus comes as no surprise that, about sixty years later, Abū Kāmil rose up in vigorous defence of al-Khwārizmī's precedence when it was contested by the grandson of ʿAbd al-Ḥamīd. In his work on the application of algebra to the partition of inheritances, he tells us the following[51]: in a book on algebra of mine, "I have established the facts concerning the precedence and the priority for the algebra of Muḥammad ibn Mūsā (al-Khwārizmī) and concerning the attribution made by Muḥarriq, known as Abū Burda, to ʿAbd al-Ḥamīd—whom he declares to be his grandfather—by showing the deficiency and the limits of his knowledge in what he attributes to his grandfather." We do not know the book on algebra in question, but in the one preserved the priority of al-Khwārizmī is reaffirmed (see below, page 63).

Applications

We have mentioned above that the study of the six equations was followed by about forty problems showing how to reduce a given problem to one of these canonical equations. Here is one example. [Appendix C.1]

I have divided ten into two parts; then I have divided this one by that one and that one by this one, which gave two dirhams and one sixth.[52]

We are thus asked to solve

$$\begin{cases} u + v = 10 \\ \frac{u}{v} + \frac{v}{u} = 2 + \frac{1}{6}. \end{cases}$$

Set $v = x$. As $u = 10 - x$, the second equation becomes

$$\frac{10-x}{x} + \frac{x}{10-x} = 2 + \frac{1}{6},$$

then

$$(10-x)^2 + x^2 = x(10-x)\left(2 + \frac{1}{6}\right),$$

yielding the equation

$$100 + 2x^2 - 20x = \left(21 + \frac{2}{3}\right)x - \left(2 + \frac{1}{6}\right)x^2.$$

[51] Extract reproduced in Vol. V, p. 68 of the bibliographical work of Hājjī Khalīfa (ca. 1650), published (together with a Latin translation) by G. Flügel, *Lexicon bibliographicum et encyclopaedicum* (7 vol.), Leipzig and London 1835–58.

[52] Edition by Rosen (note 49), pp. 44–46. The *dirham* in question is a monetary unit, the name of which is derived from the Ancient Greek drachma.

Next, by "restoration" (*al-jabr*, see page 34), we obtain the desired standard form

$$100 + \left(4 + \frac{1}{6}\right) x^2 = \left(41 + \frac{2}{3}\right) x.$$

Then, by "reduction" (*al-radd*; also called *al-ikmāl*, "completion"), that is, by dividing the equation by the coefficient of x^2, we obtain

$$x^2 + 24 = 10x,$$

the (lesser) solution of which is $5 - \sqrt{25 - 24} = 4$.

Remember that this was all expressed in words (see page 55), and so the original text looks quite different, even though the logical steps are the same. For example, the original text explains these last transformations as follows. "This becomes twenty-one things and two thirds of a thing minus two squares and one sixth (of a square), equal to one hundred and two squares minus twenty things. Restore this (equation), and (then) add two squares and a sixth (of a square) to one hundred and two squares minus twenty things, and add the twenty things subtracted from the one hundred and the two squares to the twenty-one things and two thirds of a thing. You will thus have one hundred and four squares and a sixth of a square, equal to forty-one things and two thirds of a thing. Reduce this to one square (...)."

Problems about splitting a given number into two parts that satisfy an additional condition, such as the above example, date back to antiquity. As the given number was most often taken to be ten in Muslim countries, this category of problem became known as *masā'il al-'asharāt*, that is, "problems of the tens" or, as we shall call them from now on, "ten problems." Their general form,

$$\begin{cases} u + v = 10 \\ f(u, v) = k \end{cases} \quad \text{for a fixed } k,$$

may seem banal. Their importance in the history of algebra, however, is considerable: by varying the condition in the second equation, one can complicate the form of the solution equation or the final result at will. With the condition $\frac{u}{v} + \frac{v}{u} = k$ as above, the problem becomes

$$\frac{u}{10 - u} + \frac{10 - u}{u} = k;$$

thus

$$u^2 + (10 - u)^2 = ku(10 - u),$$

or

$$(k + 2)u^2 + 100 = 10(k + 2)u,$$

which has solutions

$$u, v = \frac{5(k + 2) \pm 5\sqrt{k^2 - 4}}{k + 2}.$$

Depending on the choice of k, the solutions may be integral, rational (fractional), irrational, or even complex. With his choice of $2 + \frac{1}{6}$ for k, al-Khwārizmī obtains an integral solution. His successor Abū Kāmil chooses $k = 4 + \frac{1}{4}$ to obtain the solutions 2 and 8, then $k = \sqrt{5}$ to obtain $5\sqrt{5} - 5$ and $15 - 5\sqrt{5}$. Leonardo of Pisa (Fibonacci), who had some knowledge of Arab mathematics (see page 101), also chooses $2 + \frac{1}{6}$ and $\sqrt{5}$, as well as $3 + \frac{1}{3}$, which yields the solutions $2 + \frac{1}{2}$ and $7 + \frac{1}{2}$. A later mathematician, Bahā' al-Dīn al-ʿĀmilī (1547–1622), mentions seven problems at the end of his *Essence of Arithmetic* that were considered to be unsolvable at the time. One of these is the same ten problem with $k = u$;[53] the resulting equation, $u^3 + 100 = 8u^2 + 20u$ (where all three solutions are real), was indeed still impossible to solve algebraically in the Muslim world.

Another example of al-Khwārizmī's problems, again of the same type, is

$$\begin{cases} u + v = 10 \\ 10u = v^2. \end{cases}$$

After deducing the solution equation $u^2 + 100 = 30u$, the author states only that it can be solved as for the preceding equations. It would be interesting, however, to see what he might have said about the solutions $u = 15 \pm \sqrt{125}$, as they are irrational and only a few equations, moreover all of which are of the type $ax^2 = c$, end with an irrational solution in al-Khwārizmī's work.

3.3. Abū Kāmil

Abū Kāmil's *Algebra*

The next Arab algebraist after al-Khwārizmī is, according to the fourteenth-century encyclopedist Ibn Khaldūn, the Egyptian Abū Kāmil (ca. 850–ca. 930).[54] Although he followed al-Khwārizmī by more than two generations and did not know him personally, Abū Kāmil recognized his predecessor's role as an innovator and praised him in the introduction to his own book on algebra: "I have studied with great attention the writings of the mathematicians, examined their assertions, and scrutinized what they explain in their works; I thus observed that the book by Muḥammed ibn Mūsā al-Khwārizmī known as *Algebra*[55] is superior in the accuracy of its principles and the exactness of its argumentation. It thus behooves us, the community of mathematicians, to recognize his priority and to admit his knowledge and his superiority, as in writing his book on algebra he was an initiator and the discoverer of its principles, by which God has given us access to all that had remained obscure, has put within our reach all that had been left in darkness, has made easy all that had been arduous, and has elucidated

[53] G. Nesselmann, *Beha-eddin's Essenz der Rechenkunst*, Berlin 1843, p. 56.

[54] See Vol. III, pp. 136–37 of *Les Prolégomènes d'Ibn Khaldoun*, translated by M. de Slane (3 vol.), Paris 1863–68.

[55] Arabic: *al-jabr wa'l-muqābala*.

all that had been uncertain." Since al-Khwārizmī had opened such a royal road, one might even come to doubt the necessity of a new work on algebra. But Abū Kāmil, without tempering his praise, proceeds to reveal some imperfections in the work of his predecessor. For he continues as follows: "In it I have noticed some problems that he failed to explain and clarify. It is by the favor granted to me by the great and mighty God and by the blessings, the kindness, and the beneficence with which He has endowed me that I was given to know the science of calculation, to discover its secrets, and to elucidate its mysteries, and that appeared to me principles devoid of obscurity and arguments in no way inaccessible. I have deduced numerous problems, most of which lead to types (of equation) other than the six explained by al-Khwārizmī in his book. (...) I have thus composed a book on algebra. I explain in it part of what was explained by Muḥammed ibn Mūsā al-Khwārizmī in his book in order that mine make it possible to dispense with his; I clarified his discussion and filled gaps in al-Khwārizmī's explanations and argumentation."

One might imagine that Abū Kāmil only raised his predecessor to such heights initially in order to later accelerate his fall. Earlier, however, he had come vigorously to al-Khwārizmī's defence (see page 61). As a matter of fact, one might simply see in this introduction and his attachment to the memory of al-Khwārizmī an *oratio pro domo*: is it not clever to praise a previous work highly before asserting that one's own work is an improvement? But, in fact, there was no real need for Abū Kāmil to resort to such self-promotion, as his *Algebra* is indeed indisputably at a higher level than that of his predecessor. In any event, the readership of the two books was not meant to be the same. While al-Khwārizmī's *Algebra* addressed the general public, Abū Kāmil's was intended for mathematicians, that is to say, readers familiar with Euclid's *Elements*. Thus, even though he begins as his predecessor did, with solving the six equations and teaching algebraic calculation (Book I) and then continuing with application problems (Book II), the whole stands at a much higher level. In addition, the study of equations involving irrational numbers (Book III) and the calculations concerning polygons (Book IV) are new, as are the indeterminate problems (Book V; see above, pages 46–49); so also are the various applications of algebra to daily life (realistic situations) or to more recreational questions (unrealistic situations) with which he completes his work and which will form, from then on, the concluding part of most algebra books to early modern times.[56]

[56]Abū Kāmil's *Algebra* is preserved in Arabic in a single manuscript, in excellent condition, held in Istanbul (manuscript Kara Mustafa Paşa 379; now Bayazıt 19046). It was published as a facsimile edition by the Frankfurt Institut für Geschichte der arabisch-islamischen Wissenschaften in 1986. The *Algebra* was translated into Latin in the fourteenth century, although incompletely (it lacks the introduction and ends at the beginning of Book V); this translation was published in *Vestigia mathematica*, Amsterdam 1993, pp. 215–252 (by J. Lorch, Book IV) and pp. 315–452 (by J. Sesiano, Books I–III and the fragment of Book V). Finally, the *Algebra* was translated into Hebrew in the fifteenth century, and a portion of this (Books I–III) was published by M. Levey in *The Algebra of Abū Kāmil*, Madison

Quadratic Equations

Since knowledge of the *Elements* was assumed, the geometric representation of the solution of quadratic equations could be simplified and unified. This is achieved via two propositions from Euclid's *Elements*; both belong to Book II, the main aim of which is to prove geometrically what we think of as algebraic identities. The two propositions in question are the following.

- Proposition II.5: *If a line segment is cut into (two) equal (parts) and into (two) unequal (parts), then the rectangle formed by the unequal parts of the entire segment, along with the square on the segment between the points of section, equals the square on the half (of the entire segment).*

  That is, if AB is divided equally at F and unequally at D (see Figure 12), we have

$$\mathrm{AD} \cdot \mathrm{DB} + \mathrm{FD}^2 = \mathrm{AF}^2.$$

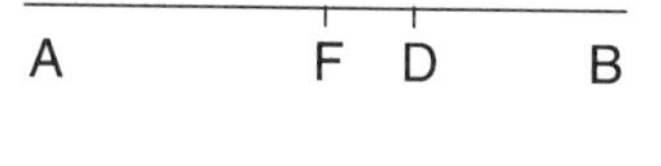

FIGURE 12

- Proposition II.6: *If a line segment is cut in half and some segment is added to it in a straight line, then the rectangle formed by the entire segment, with the added segment, and the added segment, along with the square on the half (of the initial segment), equals the square on the sum of the half and the added segment.*

  If AB is divided equally at F and DB is the extension (see Figure 13), we have

$$\mathrm{AD} \cdot \mathrm{DB} + \mathrm{FB}^2 = \mathrm{FD}^2.$$

A F B D

FIGURE 13

With AD $= u$ and DB $= v$, these two theorems become, respectively,

$$u \cdot v + \left(\frac{u+v}{2} - v\right)^2 = \left(\frac{u+v}{2}\right)^2$$

$$u \cdot v + \left(\frac{u-v}{2}\right)^2 = \left(\frac{u-v}{2} + v\right)^2,$$

1966. These texts are of uneven quality: while the Arabic manuscript is, as we said, in excellent condition, the Latin translation is good only in the mathematical sections (see page 94); the Hebrew translation is of lesser quality, and, furthermore, its modern edition is unsatisfactory.

that is, the well-known identity

$$u \cdot v + \left(\frac{u-v}{2}\right)^2 = \left(\frac{u+v}{2}\right)^2.$$

As we have said, these two propositions from Euclid allow one to considerably simplify the representations of the solutions of the quadratic equation. Suppose that, in the three cases, AC $= x^2$ and AB $= p$, and that F is the midpoint of AB.

Case 1: $x^2 + px = q$ (Figure 14)

Then we have BC $= q$. From Proposition II.6, AD$\cdot$DB$+$FB$^2 =$ FD^2, so $q + \left(\frac{p}{2}\right)^2 = \left(x + \frac{p}{2}\right)^2$, yielding the formula.

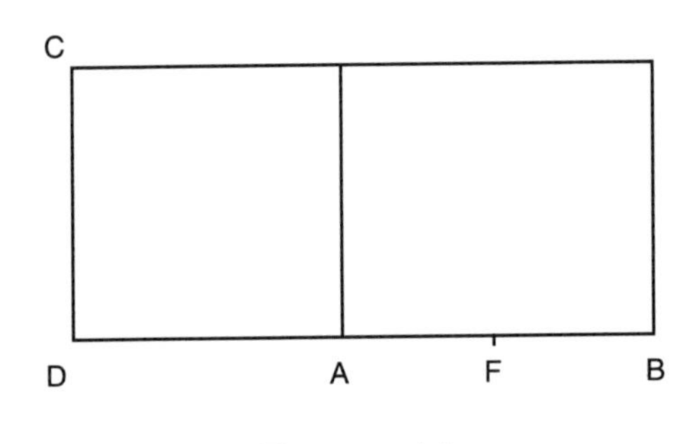

FIGURE 14

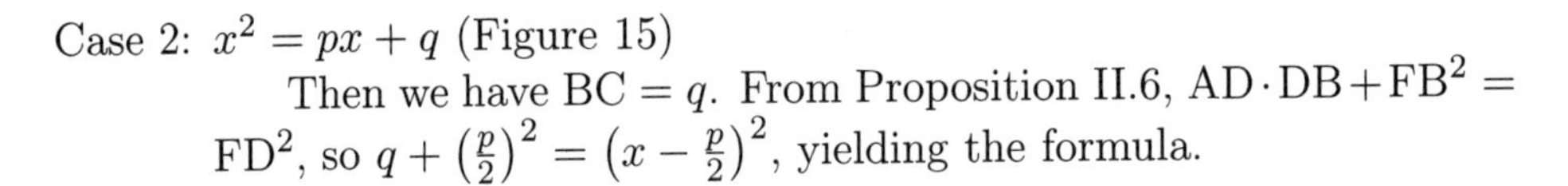

Case 2: $x^2 = px + q$ (Figure 15)

Then we have BC $= q$. From Proposition II.6, AD$\cdot$DB$+$FB$^2 =$ FD^2, so $q + \left(\frac{p}{2}\right)^2 = \left(x - \frac{p}{2}\right)^2$, yielding the formula.

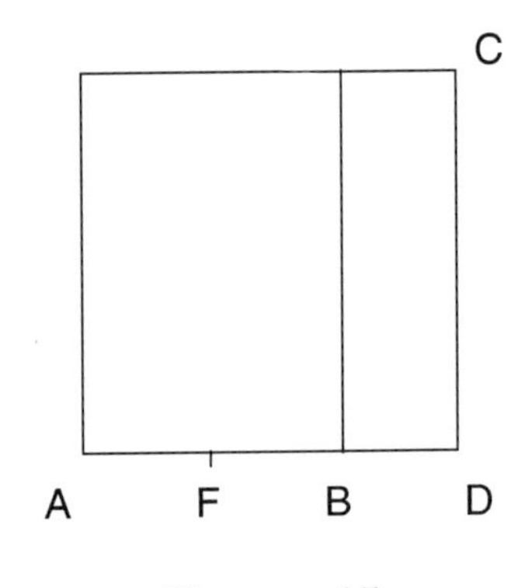

FIGURE 15

Case 3: $x^2 + q = px$ (Figures 16 and 17)

Then we have BC $= q$. From Proposition II.5, AD$\cdot$DB$+$FD$^2 =$ AF^2, so $q + \left(x - \frac{p}{2}\right)^2 = \left(\frac{p}{2}\right)^2$. But we can take the segment FD to be either $x - \frac{p}{2}$ or $\frac{p}{2} - x$, unless F and D coincide.

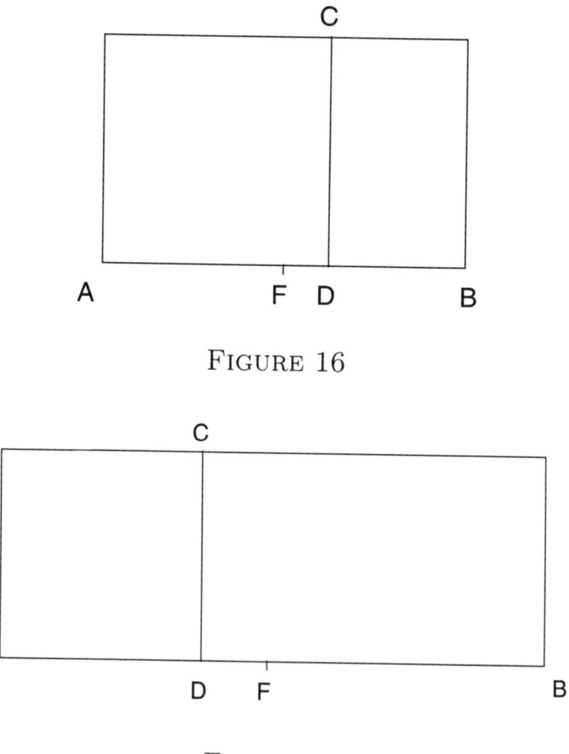

FIGURE 16

FIGURE 17

Examples of Problems

We will now give examples of some problems to illustrate two features of Abū Kāmil's *Algebra*, which are also characteristic of medieval algebra.

Example 1. We have noted the importance of geometric representations in the study of quadratic equations in Islamic countries, which was doubtlessly the result of the importance of geometry in Greece. This requirement of a *more geometrico* justification of algebraic reasoning, deemed to be necessary in a branch of mathematics that had not yet gained its autonomy, was taken so far that problems often included both an algebraic solution and its derivation from a geometric figure. This is the case, for example, in the following problem from Book II of Abū Kāmil's *Algebra*, which concerns two men who are said to have acquired ten garments for 72 dirhams, each paying the same price, 36 dirhams, while each garment for one man costs three dirhams less than each for his companion. In equations, we can write this as

$$\begin{cases} u + v = 10 \\ pu = (p - 3)\, v = 36, \end{cases}$$

which is one more example of a ten problem.

Here the geometric illustration precedes the algebraic solution. We represent the ten garments by AB, the garments of the first man by AG, and

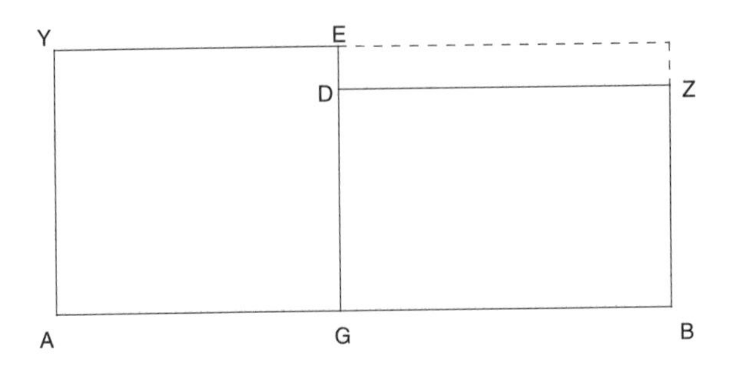

FIGURE 18

the garments of the second by GB (see Figure 18). In addition, we represent the price of one of the garments in AG by GE and the price of one of the garments in GB by GD. We thus know that DE = 3 and that AE and DB are each 36. Letting GB be the unknown x, we have that EZ $= 3x$; then BY $= 72+3x =$ AY $\cdot$ AB, and, since AB $= 10$, it follows that AY $= \frac{36}{5} + \frac{3}{10}x$. Since AY $=$ GD$+3$, GD $= \frac{21}{5} + \frac{3}{10}x$, and so DB $=$ GD$\cdot$GB $= \frac{21}{5}x + \frac{3}{10}x^2 = 36$. This yields the equation $x^2 + 14x = 120$ with solution $x = 6$.

The purely algebraic solution comes next. With x and $10 - x$ being the parts and $p - 3$ and p their respective prices, we have

$$p\,(10 - x) + (p - 3)\,x = 72$$
$$10p = 3x + 72$$
$$p = \frac{3}{10}x + \frac{36}{5}.$$

Then, as $(p - 3)x = 36$,

$$\frac{3}{10}x^2 + \frac{21}{5}x = 36,$$

which is the same equation as found above.

Of course, the solution is written purely verbally. The text reads as follows (see Appendix D, the reproduction of the Hebrew translation).

If it is said to you: Ten garments between two men with a payment of seventy-two dirhams, and each of them spent thirty-six dirhams. One of the men took some garments and the other the remainder of the garments. The price of each garment of one man is three dirhams more than the price of each garment of the other.

Its treatment consists in putting as the ten garments line AB, *the garments taken by one of them, line* AG, *the others, line* GB. *We put as the price of each garment of line* GB *line* GD, *and the whole price of the garments of line* GB *will be area* DB; *and area* DB *is thirty-six. You put as the price of each garment of line* AG *line* GE; *the whole price of the garments of line* GA *will be area* AE, *which is thirty-six. Line* DE *is three on account of its being the excess of the price of each garment of one of them over the price of each garment of the other. We put as line* GB *a thing. It is the same as line* DZ, *and line* DE *is three; therefore area* EZ *will be three things. The whole area* BY *is* 72 *and three things, and line* AB *is ten, so line* AY *is seven and a fifth and three tenths of a thing. But it is the same as line* GE. *Line* ED *of it being three, the remaining line* GD *is four and a fifth and three tenths of a thing. We multiply it by line* GB, *which is a thing. This produces four things and a fifth of a thing and three tenths of a square, equal to thirty-six dirhams. You complete three tenths of a square—its completion* (Arabic: (*al-*)*ikmāl*, see p. 62) *meaning that the square will become a whole* (*one*)*; that is, you will multiply it by three and a third, and you* (*also*) *multiply everything* (*else*) *you have by three and a third. It produces a square and fourteen things equal to a hundred and twenty dirhams. Proceed as I have told you.*[57] *The thing will amount to six, and such is the number of the garments belonging to one of them, and this is line* GB, *the other part being the remainder of ten garments, thus four, and this is line* AG. *This is what we intended to show.*

There is for this and similar problems another way, following the manner I have explained to you.[58] *It consists in putting as the garments taken by one of them a thing and as the others, ten minus a thing. You already know*[59] *that when we multiply a thing by the price of each garment of one part, it produces* 36 *dirhams and when we multiply ten minus a thing by the price of each garment of the other part it produces likewise* 36 *dirhams. And if you multiply all the garments by the price of each of the garments of a thing and three*[60] *it produces seventy-two dirhams and three things. When you divide them by the ten, it gives, as the price of each of the garments of ten minus a thing, seven and a fifth and three tenths of a thing. And the price of each of the garments of a thing will be four and a fifth and three tenths of a thing. You proceed as I told you in the foregoing.*[61] *The thing will amount to six garments, and this is what one of them took, and the other took the remainder of the ten, thus four garments.*

This problem is like the problem stating: We have divided ten into two parts; we have multiplied one part by a thing and it produces thirty-six, and we have multiplied the other part by a thing and three and it produces thirty-six.[62] *You will proceed as I told you before.*

Once again: For us the algebraic solution would suffice. Here, though, the geometric treatment helps to visualize the solution.

Example 2. In his introduction, Abū Kāmil announced that he would expound problems leading to types of equations other than the usual six types

[57]How to solve quadratic equations has been taught earlier (in Book I).
[58]Most of the previous examples are ten problems and one of the parts is taken to be x.
[59]from the data.
[60]that is: by the price, increased by 3, of each of the garments of a thing.
[61]We arrived at the same expression in the geometric solution.
[62]Here the "thing" is the lesser price.

(see above, page 64). This is the purpose of Book III of his *Algebra*, which begins as follows: "I have discovered numerous problems of algebraic calculation leading to types other than the six explained in this work. I will expound some of these, which will allow one to understand this type of algebraic calculation." The following example from Book III will show what is distinctive about this new type of equation and calculation. [Appendix C.2]

As above, this is a ten problem, in which the two quantities to be determined, u and v $(u > v)$, satisfy

$$\begin{cases} u + v = 10 \\ \left(\frac{u}{v}\right)^2 - \left(\frac{v}{u}\right)^2 = 2. \end{cases}$$

Following Abū Kāmil, we start by considering the second equation. Letting $\frac{v}{u}$ be the unknown x, this becomes

$$\frac{1}{x^2} = x^2 + 2,$$

or

$$x^4 + 2x^2 = 1,$$

so that

$$x = \sqrt{\sqrt{2} - 1} = \frac{v}{u}.$$

Then, from the first equation,

$$v = 10 - u.$$

Thus, from what we have found above,

$$\frac{10 - u}{u} = \sqrt{\sqrt{2} - 1}.$$

We then obtain

$$\frac{u^2 + 100 - 20u}{u^2} = \sqrt{2} - 1.$$

Equivalently,

$$\left(2 - \sqrt{2}\right) u^2 + 100 = 20u,$$

from which we obtain the standard form

$$u^2 + \left(100 + \sqrt{5000}\right) = u\left(20 + \sqrt{200}\right).$$

This yields the solution

$$u = 10 + \sqrt{50} - \sqrt{50 + \sqrt{5000}}.$$

What new knowledge does this treatment contain? In terms of mathematical operations, there is nothing new. In particular the relation

$$\frac{1}{\sqrt{a} \pm \sqrt{b}} = \frac{\sqrt{a} \mp \sqrt{b}}{a - b},$$

which allows a sum or difference of two square roots in the denominator of a fraction to be eliminated, was known from Book X of Euclid's *Elements*, which investigates certain irrational line segments. What is new, however, is the use of irrational quantities throughout Abū Kāmil's Book III, both as given values and—see the example above—as answers, whereas earlier authors had avoided such numbers as much as possible.

Book III comprises about thirty problems, of which we just examined the thirteenth. Some lead to much more cumbersome calculations and answers, such as the twenty-fifth, where one obtains the equation

$$x^2 + \left(176 + \frac{36}{39} + \sqrt{31\,558 + \frac{282}{1521}}\right) = \left(35 + \frac{15}{39} + \sqrt{1262 + \frac{498}{1521}}\right) x,$$

for which one positive solution is

$$x = 17 + \frac{27}{39} + \sqrt{315 + \frac{885}{1521}} + \sqrt{451 + \frac{1029}{1521} + \sqrt{395\,130 + \frac{2\,057\,670}{2\,313\,441}} - \sqrt{31\,558 + \frac{282}{1521}}}.$$

It could be said that it was with Abū Kāmil that irrational numbers truly became accepted as numbers.[63] However, such an assertion would need to be qualified somewhat. First, the most conspicuous result of this innovation was to complicate calculations, and, for that reason alone, it was destined to be received half-heartedly. In the absence of algebraic symbolism, the numerical treatment became almost incomprehensible. For example, consider simple expressions like

$$\sqrt{\alpha + \sqrt{\beta}} \qquad \text{and} \qquad \sqrt{\alpha} + \sqrt{\beta}.$$

They would both be expressed, in Arabic and in Latin, as "the root of α and the root of β." The declension in Latin and the explicit writing of vowel-marks in Arabic—the two scientific languages of the time—may, however, allow one to distinguish between them. One might also use expressions like "the sum of α and the root of β, the root of the sum being taken" for the first (note 180) and "the sum of the roots of α and β" for the second.

[63]Of course, they had already been encountered before, with the Greeks proving their existence by showing that $\sqrt{2}$, as well as other numbers, could not be represented as a quotient of two positive integers. The proof of the irrationality of $\sqrt{2}$, given by Aristotle in his *Prior Analytics* (I, 23, 41*a*), which surely dates back to the Pythagoreans, is well known: supposing that $\sqrt{2} = \frac{p}{q}$ with p and q relatively prime positive integers (one can always assume that the fraction is reduced), we have $2q^2 = p^2$; so p^2 is even and thus p is as well. Then $p = 2p'$ and $q^2 = 2p'^2$, so q^2 must be even and q must be as well, contrary to the hypothesis of relative primality. As the reasoning is correct, the initial hypothesis—that $\sqrt{2}$ can be represented as a quotient of two integers—must be false.

But the question of the span of the radical sign becomes problematic with more elaborate expressions involving more than two roots. The difficulties that the Latin translator encountered in Book III illustrate the problems caused by this lack, as is seen from his autograph copy, which is preserved; his hesitations also make evident the perplexity Arab readers must have felt when faced with Abū Kāmil's text. Thus his new types of equations probably garnered only a tepid reception, and his initiative does not appear to have been much followed. However, be it through Abū Kāmil's work or simply through the development of numerical algebra, the presence of irrational numbers in problems became more common from then on and was restricted only by the complications they introduced in calculations.

Another limitation of Abū Kāmil's innovation is that his irrational numbers are in fact just square roots. That is, the irrational numbers he considers are precisely those numbers that can be constructed by means of classical Greek geometry and that Euclid had already studied and classified, except that in the *Elements* they take the form of line segments.

By using compass and straightedge, the only two instruments allowed in geometric constructions in Euclid's *Elements*, one can draw a straight line between two given points and a circle of arbitrary radius having a given point as center (Postulates I and III in the *Elements*; Postulate II states that a line segment can be extended indefinitely, thus asserting the infinite extent of space).

In addition to drawing a line segment and a circle, these two instruments allow one to perform a number of constructions, corresponding to the four arithmetical operations. Given two segments a and b, one can

- add the two segments a and b,
- subtract the lesser of the two segments from the greater,
- multiply the two segments, and
- divide one of the segments by the other.

For the last two operations, consider an angle formed by a and b and draw the unit segment on b (see Figure 19). As

$$\frac{a}{1} = \frac{ab}{b},$$

ab is determined on the extension of a by the straight line passing through the endpoint of b and parallel to the segment connecting the endpoints of a and the unit segment. Similarly, as

$$\frac{a}{b} = \frac{\frac{a}{b}}{1},$$

$\frac{a}{b}$ is determined on a by the segment drawn from the endpoint of the unit segment and parallel to the segment joining the endpoints of a and b.

Thus, starting from the unit segment, we can construct all positive integers and extend this domain to that of all positive rational numbers. There is also an operation that allows certain irrational numbers to be constructed:

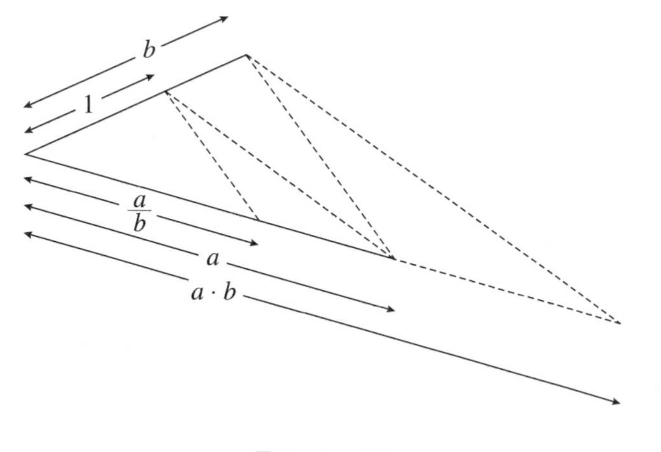

FIGURE 19

- the root of a is constructible.

This is achieved, as shown in *Elements* II.14, by adding to a the unit segment and drawing the semi-circle having this length as its diameter (see Figure 20). The height at the endpoint of a is then $\sqrt{a}$, because the height of a right triangle inscribed in a semi-circle is the geometric mean of the two segments it defines on the diameter.

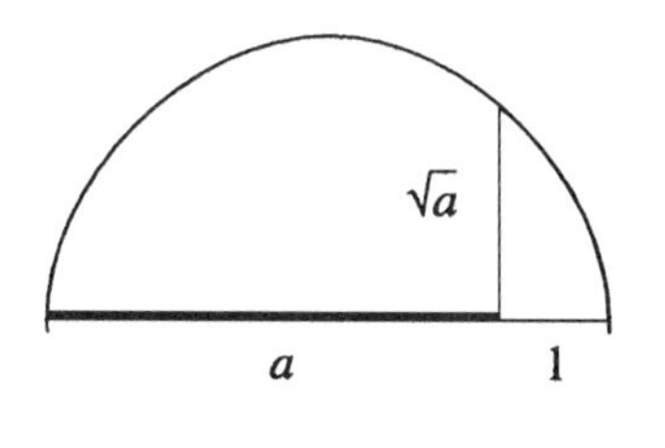

FIGURE 20

We can thus use compass and straightedge to construct any segment resulting from known segments via the operations of addition, subtraction, multiplication, division, and the extraction of square roots. This means that we can geometrically solve any problem (having a positive answer) that can be algebraically reduced to solving linear and quadratic equations. For example, Euclid can show how to construct a pentagon inscribed in a given circle using compass and straightedge; this is so because the side of a regular pentagon is related to the radius of the circumscribed circle by the relation $x = \frac{r}{2}\sqrt{10 - 2\sqrt{5}}$, expressing x as a function of r by the above operations only. Now Book IV of Abū Kāmil's *Algebra*, which precedes the section on indeterminate equations, solves algebraic problems concerning regular polygons constructible by compass and straightedge. This falls in the domain of algebra of the two lowest degrees, with the inconvenience that one must use roots. The purpose of Book III was to familiarize readers with their use.

Example 3. *An equilateral and equiangular* (that is, regular) *pentagon has an area of* 50 *(square) cubits.*[64] *What is each side of it?* (Problem IV.16)

Let $d = 2r$ be the diameter of the circle circumscribing the pentagon. As the aim of a preceding problem in the text was to find the diameter from the side of the pentagon inscribed, Abū Kāmil begins by recalling this result: "It is clear from what we have explained that if you want to know the diameter of a circle circumscribing a given equilateral and equiangular pentagon, you are to multiply one of its sides by itself, double the result, and keep the result in mind[65]; then you multiply again one of its sides by itself, then the result by itself, then you take four-fifths of this. You take the root of that and add the result to what you have kept in mind. Taking the root of this gives the diameter of the circle."

With x being the side, this means that

$$d = \sqrt{2x^2 + \sqrt{\frac{4}{5}x^4}} \quad \left(= x\sqrt{2 + \frac{2}{\sqrt{5}}} \right).^{66}$$

Now the pentagon can be decomposed into five isosceles triangles of base x and equal sides r, with height

$$h = \sqrt{r^2 - \frac{x^2}{4}}.$$

By applying the preceding relation for $r = \frac{d}{2}$, we find that this equals

$$\begin{aligned} h &= \sqrt{\frac{x^2}{4}\left(2 + \frac{2}{\sqrt{5}}\right) - \frac{x^2}{4}} \\ &= \sqrt{\frac{x^2}{4}\left(1 + \frac{2}{\sqrt{5}}\right)} \\ &= \frac{x}{2}\sqrt{\frac{2+\sqrt{5}}{\sqrt{5}}}. \end{aligned}$$

Since 50 is the area of the whole pentagon, then

$$\frac{1}{2}xh = \frac{x^2}{4}\sqrt{\frac{2+\sqrt{5}}{\sqrt{5}}} = 10,$$

so

$$\frac{x^4}{16}\left(\frac{2+\sqrt{5}}{\sqrt{5}}\right) = 100.$$

---

[64] Here is another example of a length unit taken as area unit (note 24).

[65] As remarked above (note 12), this means that we have completed an intermediate calculation.

[66] As $x = \frac{r}{2}\sqrt{10 - 2\sqrt{5}}$, we then have that $2r = \frac{4x}{\sqrt{10-2\sqrt{5}}} = \frac{4x\sqrt{10+2\sqrt{5}}}{\sqrt{80}} = x\sqrt{2 + \frac{2}{\sqrt{5}}}$.

Hence

$$x^4\left(\frac{2+\sqrt{5}}{\sqrt{5}}\right)=1600.$$

Abū Kāmil then summarizes the remainder of the solution, as follows: "Reduce everything you have to x^4 by multiplying all by five less the root of twenty;"

$$\left[\text{for } x^4\left(\frac{2+\sqrt{5}}{\sqrt{5}}\right)=x^4\frac{(\sqrt{5}+2)(\sqrt{5}-2)}{\sqrt{5}(\sqrt{5}-2)}=x^4\frac{1}{5-\sqrt{20}}\right]$$

"then x^4 will equal eight thousand minus the root of fifty-one million two hundred thousand. Take the root of this (twice); it will give each side of the pentagon."

$$\left[\text{Since } x^4=1600\left(5-\sqrt{20}\right), \text{ so } x=\sqrt[4]{8000-\sqrt{51\,200\,000}}\right].$$

Buying Birds

In addition to his *Algebra* (and other minor treatises), Abū Kāmil wrote a short book teaching how to solve indeterminate linear systems with (positive) integral solutions.[67] These problems were known in the East as "bird problems" (*masāʾil al-ṭuyūr* in Arabic), as they often involved the purchasing of various types of bird, given the total value of the purchase, the total number of birds, and the price per bird of each species. In our notation, if x_i is the (integral) number of birds of the ith species, p_i the price of one of them, and k the number of birds to be bought with sum l, we have

$$\begin{cases}\displaystyle\sum_{i=1}^{n} x_i=k\\ \displaystyle\sum_{i=1}^{n} p_i x_i=l.\end{cases}$$

It appears that what aroused Abū Kāmil's interest in such systems was the great variety in the number of solutions they could yield. The first five examples he considers involve pairs of equations with three to five unknowns; he finds, respectively, 1, 6, 96 (although, in fact, there are 98), 304, and 0 solutions. But this all seems banal compared to the last problem, which, according to the end of the introduction, moved Abū Kāmil to write his book: "I found myself before a problem that I solved and for which I discovered a great many solutions; looking deeper for its solutions, I obtained two thousand six hundred and seventy-six correct ones. My astonishment about that was great, but I found out that, when I recounted this discovery,

[67] H. Suter, "Das Buch der Seltenheiten der Rechenkunst von Abū Kāmil el-Miṣrī," *Bibliotheca Mathematica*, 3. F., 11 (1910–11), pp. 100–120 (reprinted by the Frankfurt Institut für Geschichte der arabisch-islamischen Wissenschaften in 1986 in H. Suter, *Beiträge zur Geschichte der Mathematik und Astronomie im Islam*, 2 vol.).

those who did not know me were arrogant, shocked, and suspicious of me. I thus decided to write a book on this kind of calculation, with the purpose of facilitating its treatment and making it more accessible."

In this sixth problem, one must buy a hundred birds with a hundred dirhams (these givens are common to the six problems); the ducks cost two dirhams each, while one dirham buys two pigeons, three woodpigeons, four larks, or one chicken. In equations, this means that

$$\begin{cases} x_1 + x_2 + x_3 + x_4 + x_5 = 100 \\ 2x_1 + \frac{1}{2}x_2 + \frac{1}{3}x_3 + \frac{1}{4}x_4 + x_5 = 100. \end{cases}$$

Abū Kāmil uses particular names for each of four unknowns: *shay'* ("thing"), the usual name for the unknown, for x_1; the two monetary units *dīnār* and *fals* for x_2 and x_3; and *khātam* ("seal") for x_4.[68]

To solve this system, Abū Kāmil begins by setting equal the two expressions for x_5:

$$100 - x_1 - x_2 - x_3 - x_4 = 100 - 2x_1 - \frac{1}{2}x_2 - \frac{1}{3}x_3 - \frac{1}{4}x_4,$$

which implies that

$$(*) \qquad x_1 = \frac{1}{2}x_2 + \frac{2}{3}x_3 + \frac{3}{4}x_4.$$

Then, since x_5 must be strictly positive (solutions equal to zero are not allowed), we obtain

$$x_1 + x_2 + x_3 + x_4 = \frac{3}{2}x_2 + \frac{5}{3}x_3 + \frac{7}{4}x_4 < 100.$$

Relation (*) implies that x_1 will be an integer if x_3 is divisible by 3 and either x_2 is odd and x_4 is even but not divisible by 4 or x_2 is even and x_4 is divisible by 4. We will examine each case in turn.

- First, suppose that x_2 is odd. Then

  $$\begin{aligned} x_2 &= 2k+1 \\ x_3 &= 3l \\ x_4 &= 4j+2, \end{aligned}$$

  so

  $$x_1 = k + 2l + 3j + 2$$

  with k, l, j integers, $k, j \geq 0$, $l \geq 1$. As $x_1+x_2+x_3+x_4 < 100$, we must have $3k+5l+7j < 95$, which implies that $3k+7j < 90$. Thus

[68]Note the survival in Arabic of monetary designations used in antiquity, as previously remarked about the drachma (see note 52): *dīnār* is the Latin *denarius* and *fals* is the Greek (and Latin) *follis*.

$k < 30$, $j < 13$, and (by taking $k = j = 0$) $l < 19$. The possible choices are then

$$k = 0, 1, \ldots, 29$$
$$l = 1, 2, \ldots, 18$$
$$j = 0, 1, \ldots, 12.$$

This basically follows Abū Kāmil's reasoning. The only difference is that he does not use our parameters k, l, j, but instead reasons directly with the unknowns. He thus finds the possibilities

$$x_2 = 1, 3, \ldots, 59$$
$$x_3 = 3, 6, \ldots, 51 \ (sic)$$
$$x_4 = 2, 6, \ldots, 50.$$

Abū Kāmil then turns to counting the solutions (none of the preserved manuscripts contains the table of solutions that the original text must have included in an appendix). First, he takes $x_3 = 3$ (corresponding to $l = 1$), then each of the odd values from 1 to 59 for x_2 and, for each value of x_2, all possible values of x_4 starting with 2. He thus obtains, as he notes, 212 solutions. Next, he takes $x_3 = 6$, simply remarking that one can proceed as before, and likewise for the subsequent values from $x_3 = 9$ to $x_3 = 51$. This yields, he writes, 1443 solutions. However, there are in fact 1445 solutions: Abū Kāmil omits the two solutions with $x_3 = 54$ ($x_2 = 1$, $x_1 = 38$, $x_4 = 2$, $x_5 = 5$; and $x_2 = 3$, $x_1 = 39$, $x_4 = 2$, $x_5 = 2$).

- Suppose now that x_2 is even. For

$$x_1 = \frac{1}{2}x_2 + \frac{2}{3}x_3 + \frac{3}{4}x_4$$

to be an integer, we take

$$x_2 = 2k, \qquad x_3 = 3l, \qquad x_4 = 4j,$$

thus obtaining

$$x_1 = k + 2l + 3j.$$

As $x_5 = 100 - 3k - 5l - 7j > 0$ and k, l, and j are positive integers, we obtain the conditions $3k < 88$, $5l < 90$, $7j < 92$. The possible choices are therefore

$$k = 1, \ldots, 29$$
$$l = 1, \ldots, 17$$
$$j = 1, \ldots, 13,$$

or, following Abū Kāmil,

$$x_2 = 2, \ldots, 58$$
$$x_3 = 3, \ldots, 51$$
$$x_4 = 4, \ldots, 52.$$

Abū Kāmil only briefly deals with this second case, simply mentioning the first solution ($x_2 = 2$, $x_3 = 3$, $x_4 = 4$, so $x_1 = 6$ and $x_5 = 85$) and stating that one can proceed "just as in the first case." We would thus obtain 1233 solutions (this number is not mentioned explicitly but can be deduced from the numbers 1443 and 2676 already mentioned).

We do not know where Abū Kāmil found the idea for these problems—however, one might instead ask how he could have failed to find it, as occasions to do so were certainly not lacking. Such problems, also concerning the purchasing of birds, appeared in China at the beginning of the Christian era, and then in India during the time of Abū Kāmil[69]. Moreover, similar ones were also solved in antiquity. Some were preserved in Latin translation and collected shortly before 800, along with other problems and arithmetical riddles, by the monk Alcuin, whom Charlemagne had charged with reviving education throughout the empire. The title of this collection, *Problems to stimulate acuity in the young* (*Propositiones ad acuendos iuvenes*), clearly reveals its purpose. It is highly probable that these problems originated in Greek Alexandria, as suggested by the fact that two of them involve camels, which are unlikely to have inspired a scholarly monk of Charlemagne's empire. Here are two examples of systems appearing in this collection (Alcuin's text contains only the statements and the answers).[70]

$$\begin{cases} x_1 + x_2 + x_3 = 12 \\ 2x_1 + \frac{1}{2}x_2 + \frac{1}{4}x_3 = 12 \end{cases}$$

$$\begin{cases} x_1 + x_2 + x_3 = 100 \\ 10x_1 + 5x_2 + \frac{1}{2}x_3 = 100. \end{cases}$$

In the first system, multiplying the first equation by 2 and setting equal the two expressions for $2x_1$ yields

$$24 - 2x_2 - 2x_3 = 12 - \frac{1}{2}x_2 - \frac{1}{4}x_3.$$

Thus

$$\frac{3}{2}x_2 + \frac{7}{4}x_3 = 12,$$

and so

$$6x_2 + 7x_3 = 48.$$

[69]See pp. 613–14 of J. Tropfke, *Geschichte der Elementarmathematik* (fourth edition of the sections on arithmetic and algebra, revised by K. Vogel, K. Reich, H. Gericke), Berlin 1980.

[70]*Patrologia, series latina*, ed. J. P. Migne [full title: *Patrologiæ cursus completus, sive bibliotheca* (...) *omnium SS. Patrum, doctorum scriptorumque ecclesiasticorum* (...)] (221 vol., Paris 1844-64), Vol. 101, Coll. 1143–60. More recent edition: M. Folkerts, "Die Alkuin zugeschriebenen Propositiones ad acuendos iuvenes," *Denkschriften der österreichischen Akademie der Wissenschaften, Mathem.-naturwiss. Klasse*, Vol. 116, 6, Vienna 1978.

This equation has only one acceptable solution, $x_2 = 1$, $x_3 = 6$, which yields $x_1 = 5$ (the solution $x_1 = 4$, $x_2 = 8$, $x_3 = 0$ is not allowed: the statement mentions *three* unknown quantities).

In the second system, multiplying the second equation by 2 and applying the first yields

$$200 - 20x_1 - 10x_2 = 100 - x_1 - x_2,$$

so

$$19x_1 + 9x_2 = 100.$$

Once again, there is only one acceptable solution: $x_1 = 1$, $x_2 = 9$, $x_3 = 90$.

The subject of these two problems is not the purchasing of birds. In the first, loaves of bread are to be divided among three categories of clergymen, while various kinds of pigs are to be bought in the second. The same two systems are known from medieval times in Egypt as well, where they appear in a compendium on various sciences intended for state employees.[71] But, of course, the subjects of these problems could not be the same, so the clergymen became people (men, women, children), while the pigs were replaced by oxen. [Appendices C.3 and B.7]

3.4. The Geometric Construction of Solutions of the Quadratic Equation

The geometric illustrations of the three types of quadratic equation that we have seen differ in that al-Khwārizmī's are fairly intuitive, while Abū Kāmil's, intended for people educated in mathematics, use theorems from Euclid. But mathematicians familiar with Euclid would not be satisfied with mere illustrations, and it was necessary to show how to *construct* the solution from the given values. An anonymous text dating from year 395 of the hegira (1004–05 in the Christian calendar), which, according to the author, is a compilation from various sources, provides both the illustration and the construction of the solution.[72]

To each of the three types of complete quadratic equation, the author of this text applies the so-called "application of areas" constructions taught by Euclid in his *Elements* (theorems VI.28–29). These allow one to "construct a rectangle of known area with base lying on a segment of known length (possibly extended) and differing by a square area from the rectangle of the same height having the known segment as its base." The rectangle of known area appears by rewriting the equations in terms of products:

$$\begin{aligned} x^2 + px = q &\longrightarrow x(x+p) = q \\ x^2 = px + q &\longrightarrow x(x-p) = q \\ x^2 + q = px &\longrightarrow x(p-x) = q. \end{aligned}$$

[71] Arabic manuscript 4441 of the Bibliothèque nationale de France, fol. 44^v, 6–14. This copy is dated 1571 (979 of the hegira).

[72] Manuscript 5325 of the Astan Qods Library, Mashhad.

Case 1: $x^2 + px = q$ (Figure 21)

Let AB $= p$ and CB $= \left(\frac{p}{2}\right)^2$. On the base of this square, construct the larger square CE $= \left(\frac{p}{2}\right)^2 + q$, which has known side because we can determine the square root of a known value (see page 72). The desired rectangle is then AE, and the desired solution is BD = BF. Indeed, the rectangle AE, which has known area q because it equals the sum of the three lateral areas in CE, differs from AD by a square area.

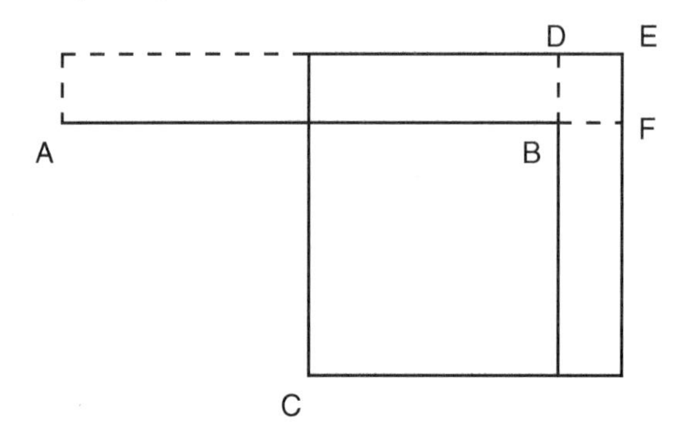

FIGURE 21

Case 2: $x^2 = px + q$

The construction is the same, but this time the solution x is the line segment AF. It is easy to check that we indeed have that AF $\cdot$ BF $= x\,(x - p) = q$.

Case 3: $x^2 + q = px$ (Figure 22)

Once again, let AB $= p$ and let CB be the square on half of AB. Square CE $= \left(\frac{p}{2}\right)^2 - q$ is now smaller than square CB, and their difference, the sum of the areas ID and DK, is AE $= q$. In this case, there are two rectangles that satisfy the condition: AE, corresponding to the solution DE = DB $= x$; and DG, which has the same size as AE, corresponding to the solution AD = DH $= x'$. Note that this case with two positive solutions reveals the relations $x + x' =$ AB $= p$ and $x \cdot x' =$ AE $= q$.[73]

3.5. The Cubic Equation

The Birth of Cubic Problems

The author of the anonymous treatise discussed above concludes by remarking that one cannot construct the solution of the cubic equation using only

[73]Today, we write $x + x' = -p$ for the first, because we consider the quadratic equation in the form $x^2 + px + q = 0$. These relations are called *Vieta's formulas*, after François Viète (1540–1603), who extended them up to the fifth degree in his *De recognitione et emendatione æquationum*.

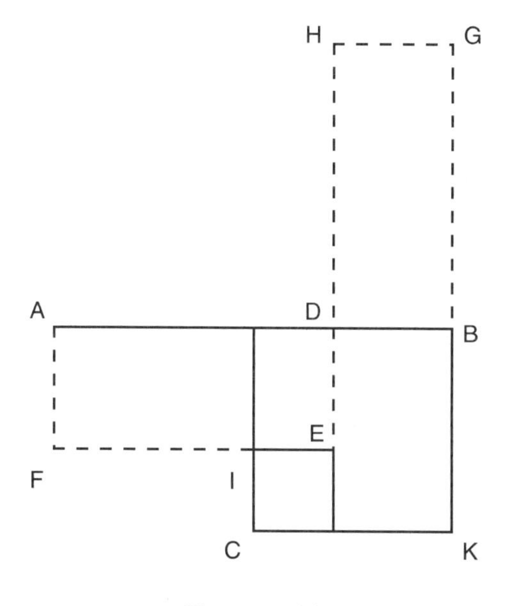

FIGURE 22

Euclid's *Elements* and that it is necessary to resort to conic sections. He mentions the various forms this equation takes when it is composed of three or four terms with positive coefficients and has, or may have, a positive solution. All these cases had perhaps already been solved at the time the treatise was written; all these solutions were in any case collected, a century later, in 'Umar Khayyām's *Algebra*, to which we will return later.

These geometric solutions of cubic equations are one of the major discoveries of Islamic mathematics. However, it was in Greece that this problem originally appeared, with the solution to the problem of doubling the cube. While a plague was ravaging Delos, we are told, the oracle there was consulted as to the means of putting an end to the epidemic by appeasing the gods. The oracle's response was to declare that the volume of Apollo's altar must be doubled while maintaining its cubic shape. The residents of Delos naively began by doubling the length of the sides, whereupon the intensity of the plague doubled as well. The gods, they were then told by a philosopher, wished only to show the Delians what unfortunate consequences neglecting the mathematical sciences could have.

Thus the Delian problem, as it has also been called since then, consists in transforming a cube of known side, and thus volume, into a cube of twice the known volume; that is, we are asked to construct the cube of volume $x^3 = 2a^3$, or its side $x = \sqrt[3]{2}a$. Because one of the factors is a cube root, this is not constructible by compass and straightedge. A transformation attributed to Hippocrates of Chios (ca. 450 B.C.) reduces this to a specific problem in plane geometry: we will find a solution if we can determine two mean proportionals between two given unequal segments, that is if, given a

and b, we can construct x and y such that

$$\frac{b}{y} = \frac{y}{x} = \frac{x}{a}.$$

Indeed, we then have

$$\frac{b}{y} \cdot \frac{y}{x} \cdot \frac{x}{a} = \left(\frac{x}{a}\right)^3,$$

and also

$$\frac{b}{y} \cdot \frac{y}{x} \cdot \frac{x}{a} = \frac{b}{a}.$$

Then, by setting equal the two right sides, we obtain $x^3 = ba^2$, and, by taking $b = 2a$, $x^3 = 2a^3$. Various geometric constructions were used in Greece to solve this problem. From Hippocrates' reduction it also follows that x (and y) can be determined by finding the point of intersection of the two curves $x^2 = ay$ and $y^2 = bx = 2ax$ or of one of these with the curve $xy = 2a^2$. According to Greek sources, this is how Menaechmus (ca. 350 B.C.) came to discover conic sections.

More than a millenium later, it was two of Archimedes' problems that served as catalyst for the study of cubic equations in the Muslim world. The first arises in Proposition II.4 of Archimedes' *On the Sphere and Cylinder*. The problem consists in cutting a sphere by a plane in such a way that the volumes of the two resulting portions of the sphere are to one another in a given ratio. Algebraically, let r be the radius of the sphere and h be the height of one of the portions (so $2r - h$ is the height of the other). Then, by the formula $V = \frac{\pi h}{6}\left(3\rho^2 + h^2\right)$, with ρ the radius of the circular base of the portion, the volumes of the two portions are, respectively,

$$V_1 = \frac{1}{3}\pi h^2 (3r - h)$$

$$V_2 = \frac{1}{3}\pi (2r - h)^2 (r + h).$$

Let α be the given (rational) ratio. We can then write the condition that $V_1 : V_2 = \alpha$ as

$$h^2 (3r - h) = \alpha (2r - h)^2 (r + h).$$

Thus

$$3rh^2 - h^3 = 4r^3\alpha + 4r^2h\alpha - 4r^2h\alpha - 4rh^2\alpha + rh^2\alpha + h^3\alpha,$$

so

$$(\alpha + 1)\, h^3 + (r\alpha - 4r\alpha - 3r)\, h^2 + 4r^3\alpha = 0,$$

and, finally,

$$h^3 - 3rh^2 + \frac{4r^3\alpha}{\alpha + 1} = 0.$$

An algebraic solution was not found for this problem, but it was solved geometrically, first by al-Khāzin (ca. 930) and then by Ibn al-Haytham (ca. 965–ca. 1040).[74]

The other seminal problem was on determining the side of a regular heptagon. In Euclid's time (300 B.C.), and surely long before, Greek mathematicians knew how to geometrically construct, by compass and straightedge, regular polygons with 2^{k+2}, $3 \cdot 2^k$, $5 \cdot 2^k$, and $15 \cdot 2^k$ (k integer ≥ 0) sides.[75] As the first two polygons which are not constructible are the heptagon and the nonagon, it is not surprising that they were studied first. To determine the side of the heptagon, Archimedes used a "mechanical" construction (the insertion of a segment of given length between two lines or curves so that its extension passes through a given point). Muslim mathematicians also attempted to geometrically construct the heptagon and the nonagon. As the degree of the algebraic relation linking the side of these two polygons and the radius of the circumscribed circle is three, they were naturally led to the study of cubic equations.

One elegant method for establishing the algebraic equation involved is a generalization of Euclid's construction for determining the side of the pentagon (*Elements* IV.10–11). Consider an isosceles triangle of base x and equal sides a, with vertex angle α and base angles $k\alpha$ (for a positive integer k). As $(2k+1)\,\alpha = 180°$, $\alpha = \frac{360°}{2(2k+1)}$. Thus if the angle α is placed at the center of a circle of radius a, then x will be the side of the inscribed regular polygon having $2\,(2k+1)$ sides. Moreover, if the same triangle is inscribed in a circle, then x will be the side of the inscribed regular polygon of $2k+1$ sides, as the angle at the center is twice that at the circumference. (The same result also follows more simply by connecting every second vertex in the $2\,(2k+1)$-gon.) We will now study these triangles, and thus also the relation between x and a.

- If $k = 1$, then the triangle with vertex at the center of the circle is equilateral. Indeed, the side of a regular hexagon equals the radius of the circumscribed circle.
- If $k = 2$, a more elaborate construction is necessary (this has been done in Figure 23, with $\mathrm{BC} = x$, $\mathrm{AB} = \mathrm{AC} = a$). With $\mathrm{AD} = x$, then $\mathrm{DC} = a - x$. As $\frac{\mathrm{AB}}{\mathrm{BC}} = \frac{\mathrm{BC}}{\mathrm{DC}}$, $\mathrm{DC} = \frac{\mathrm{BC}^2}{\mathrm{AB}} = \frac{x^2}{a}$. Thus $a - x = \frac{x^2}{a}$, whence $x^2 + ax = a^2$, which has solution $x = \frac{a}{2}\left(\sqrt{5} - 1\right)$, x then being the side of the decagon.

[74]The solution involves a parabola and a hyperbola. See pp. 91–93 in F. Woepcke, *L'Algèbre d'Omar Alkhayyâmî*, Paris 1851 (reprinted in the *Etudes*, see note 41).

[75]This does not represent all constructible regular polygons. Gauss's *Disquisitiones arithmeticae* (1801) provides the general condition: a regular polygon is constructible by compass and straightedge if the number of its sides is of the form $n = 2^t \cdot p_1 \cdot p_2 \cdot \ldots \cdot p_l$, where t is an integer ≥ 0 and the p_i's are distinct *Fermat primes* (primes of the form $2^n + 1$; only five are known: $3, 5, 17, 257, 65537$).

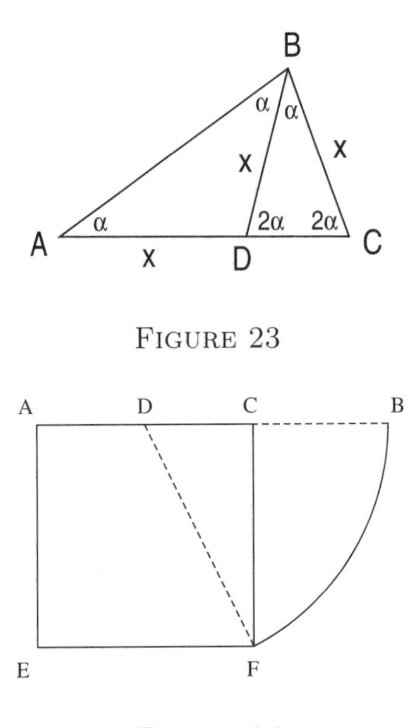

FIGURE 23

FIGURE 24

Since the equation is quadratic, we can construct x geometrically (this is the aim of the propositions from the *Elements* mentioned above). Consider Figure 24. Draw square ACFE with side a, and let D be the midpoint of AC. Connect D and F, and let B be the intersection of the extension of DC with the arc of radius DF; then $\mathrm{DB} = \mathrm{DF} = \sqrt{(\frac{a}{2})^2 + a^2} = \frac{a}{2}\sqrt{5}$. Hence, since $\mathrm{AD} = \mathrm{DC} = \frac{a}{2}$, then $\mathrm{AB} = \frac{a}{2}\left(\sqrt{5}+1\right)$ and $\mathrm{CB} = \frac{a}{2}\left(\sqrt{5}-1\right)$. Thus $\mathrm{BC} = x$ is the side of the decagon inscribed in the circle of radius $\mathrm{AC} = a$ (while DB is the side of the pentagon inscribed in the same circle). Note also that $\frac{\mathrm{AB}}{\mathrm{AC}} = \frac{\mathrm{AC}}{\mathrm{CB}}$; this is the golden ratio ($= 1.61803\ldots$).

- Let $k = 3$. Suppose that the construction has already been made, with the length x forming a zigzag line between AB and AC (see Figure 25). Draw BT and EU perpendicular to AC. Here, as in both the preceding and subsequent cases, we have $\mathrm{DC} = \frac{x^2}{a}$, for the farthest-right triangle is similar to the original triangle. Then

$$\mathrm{AD} = a - \frac{x^2}{a}$$

so

$$\mathrm{UD} = \mathrm{AU} = \frac{a}{2} - \frac{x^2}{2a}.$$

By similar triangles,

$$\frac{\mathrm{AU}}{\mathrm{AE}} = \frac{\mathrm{AT}}{\mathrm{AB}},$$

so

$$\begin{aligned}\mathrm{AB}\cdot\mathrm{AU} &= \mathrm{AE}\cdot\mathrm{AT}\\ &= \mathrm{AE}\left(\mathrm{AC}-\frac{1}{2}\mathrm{DC}\right).\end{aligned}$$

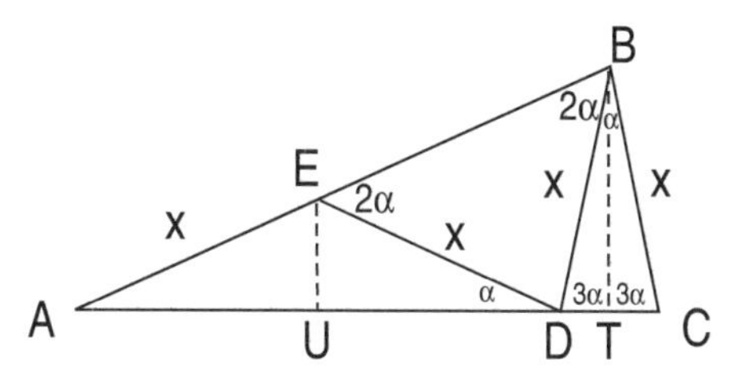

FIGURE 25

This yields the equation

$$a\left(\frac{a}{2}-\frac{x^2}{2a}\right)=x\left(a-\frac{x^2}{2a}\right).$$

Hence

$$a^3-ax^2=2a^2x-x^3,$$

so

$$x^3+a^3=ax^2+2a^2x.$$

Therefore the side of the heptagon, or of the 14-gon, cannot be constructed using only compass and straightedge. Instead, conic sections must be used, which is how mathematicians in Islamic countries in the late tenth century, such as Abū'l-Jūd and al-Sijzī, proceeded for constructing the above figure.[76]

A similar construction is found in a work of Archimedes preserved only in Arabic translation, the *Book of Lemmas*.[77] It concerns the problem of trisecting a given angle, which was as famous in antiquity as that of doubling the cube, and which leads to a cubic equation as well. In the same figure, let BDC be the angle to be trisected. Draw a circle having radius DB and center D, then determine the point E on this circle such that the extensions of lines BE and CD intersect at a point A such that AE be equal to the radius of the circle. Then BAC is a third of the given angle.

---

[76] See J. Hogendijk, "Greek and Arabic constructions of the regular heptagon," *Archive for history of exact sciences*, 30 (1984), pp. 197–330.

[77] See Vol. II, p. 518 in *Archimedis opera omnia cum commentariis Eutocii*, ed. J. Heiberg (3 vol.), 2nd ed., Leipzig 1910–15 (reprint: Stuttgart 1972).

- Now, let $k = 4$ (see Figure 26). In addition to $\mathrm{DC} = \frac{x^2}{a}$ and $\mathrm{AF} = x$, we have $\mathrm{EB} = x$, as triangle BDE is equilateral. We then know that $\mathrm{AE} = a - x$, and thus $\mathrm{AV} = \frac{a}{2} - \frac{x}{2}$. Also, by similar triangles, $\frac{\mathrm{AV}}{\mathrm{AF}} = \frac{\mathrm{AT}}{\mathrm{AB}}$. Therefore, as above,
$$\mathrm{AB} \cdot \mathrm{AV} = \mathrm{AF} \cdot \mathrm{AT} = \mathrm{AF}\left(\mathrm{AC} - \frac{1}{2}\mathrm{DC}\right),$$
and then
$$a\left(\frac{a}{2} - \frac{x}{2}\right) = x\left(a - \frac{x^2}{2a}\right).$$
Thus
$$a^3 - a^2x = 2a^2x - x^3,$$
so
$$x^3 + a^3 = 3a^2x.$$
The question of determining point E on AB, which is the key to determining the side of the nonagon and of the 18-gon, was the subject of correspondence around A.D. 1000 between two Muslim scholars, Abū'l-Jūd (mentioned in the previous case) and Bīrūnī, one of the most distinguished minds of medieval Iran (also mentioned before, page 53).[78]

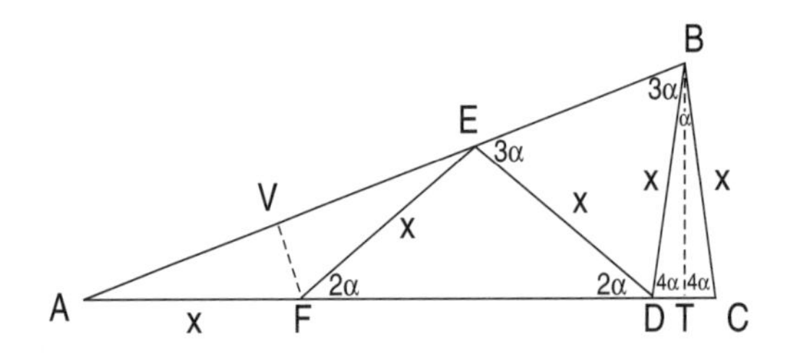

FIGURE 26

- Of course, there now arises the question of extending this method to higher polygons. We shall therefore consider what happens for $k = 5$ (see Figure 27). We have $\mathrm{DC} = \frac{x^2}{a}$, so $\mathrm{AD} = a - \frac{x^2}{a}$, and examining similar triangles shows that
$$\frac{\mathrm{AW}}{\mathrm{AD}} = \frac{\mathrm{AV}}{\mathrm{AF}} = \frac{\mathrm{AU}}{\mathrm{AG}} = \frac{\mathrm{AT}}{\mathrm{AB}} = \frac{1}{a}\left(a - \frac{x^2}{2a}\right).$$
From this sequence of equalities, we deduce that
$$\mathrm{AW} = \frac{1}{a}\left(a - \frac{x^2}{2a}\right)\left(a - \frac{x^2}{a}\right) = a - \frac{3x^2}{2a} + \frac{x^4}{2a^3},$$
$$\mathrm{AU} = \frac{x}{a}\left(a - \frac{x^2}{2a}\right),$$

[78]This correspondence is found in Woepcke, *op. cit.* (note 74), pp. 125–126.

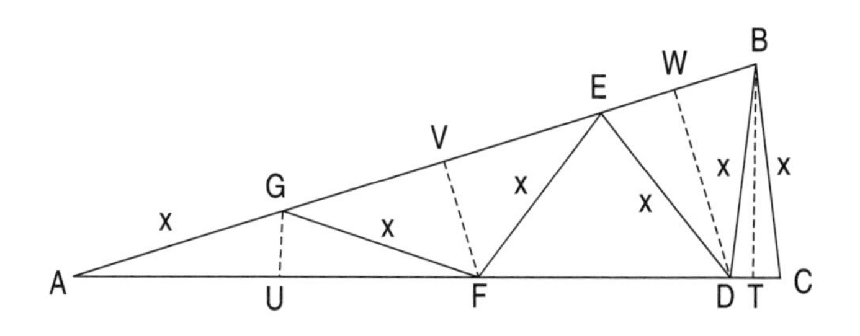

FIGURE 27

and thus, from the latter,

$$\mathrm{AF} = \frac{2x}{a}\left(a - \frac{x^2}{2a}\right).$$

From this and the initial sequence of equalities, we deduce that

$$\mathrm{AV} = \frac{2x}{a^2}\left(a - \frac{x^2}{2a}\right)^2 = 2x - \frac{2x^3}{a^2} + \frac{x^5}{2a^4},$$

and thus that

$$\mathrm{GV} = \mathrm{AV} - \mathrm{AG} = x - \frac{2x^3}{a^2} + \frac{x^5}{2a^4},$$

and so

$$\mathrm{GE} = 2\mathrm{GV} = 2x - \frac{4x^3}{a^2} + \frac{x^5}{a^4},$$

while

$$\mathrm{EB} = \mathrm{AB} - \mathrm{AG} - \mathrm{GE} = a - 3x + \frac{4x^3}{a^2} - \frac{x^5}{a^4},$$

and so

$$\mathrm{WB} = \frac{a}{2} - \frac{3x}{2} + \frac{2x^3}{a^2} - \frac{x^5}{2a^4}.$$

Therefore, we find that

$$\mathrm{AW} = \mathrm{AB} - \mathrm{WB} = \frac{a}{2} + \frac{3x}{2} - \frac{2x^3}{a^2} + \frac{x^5}{2a^4}.$$

Setting this equal to the other expression found for AW, we have

$$a - \frac{3x^2}{2a} + \frac{x^4}{2a^3} = \frac{a}{2} + \frac{3x}{2} - \frac{2x^3}{a^2} + \frac{x^5}{2a^4},$$

so, by multiplying by $2a^4$,

$$x^5 - ax^4 - 4a^2x^3 + 3a^3x^2 + 3a^4x - a^5 = 0,$$

thus a quintic equation.

Remark.

- A similar figure is found in Newton's *Arithmetica universalis* (1707, 1722; Problem XXIX in the *resolutio quæstionum*). But there a is not imposed and drawing in a given angle a succession of equal lines x determines the successive integral multiples of the given angle.

'Umar Khayyām's Solution Methods

A cubic equation has three roots, and, when the coefficients are real, then one root must be real and the remaining two are either both real or complex conjugates. Just as for the quadratic equation, in the Middle Ages the cubic equation was represented via various cases of equality between two expressions composed of positive terms only, and leading to a positive solution. The book on algebra written by the Persian mathematician 'Umar Khayyām (ca. 1048–1130) collects and completes the earlier, partial studies on these equations.

'Umar Khayyām (whose name is often transcribed as Omar Khayyam) is well known for his *Rubaiyat* (*rubā'yyāt*, "quatrains"), a collection of (at times, ribald) poems, many of which appear to have been added by others later. It is said that he was one of three young men educated together, who promised that if ever one found himself in a high position he would use it to benefit his two friends. And, indeed, one later became a vizier. He kept his word by making the third friend a chamberlain; however, the ungrateful man then attempted to supplant the vizier, was consequently replaced, and later returned to assassinate his old friend, who had recently fallen into disgrace. 'Umar Khayyām was of a different character: he asked of the vizier only the means to pursue his scientific and literary activities.

One result of the first activity is his book on algebra, which explains not only how to solve the six (linear and quadratic) equations, both numerically and geometrically, but then also how to solve cubic equations, although only geometrically. These cubic equations can be divided into fourteen types, listed in Table 1 along with the types of curves (circle, hyperbola, parabola) that 'Umar Khayyām uses to construct the positive solution.

Thus in Cases 1, 2, 4, 5, 7, 8, 11, and 12, there is always one, and exactly one, positive solution. In Cases 3, 6, 9, 10, and 14, there will be two positive solutions, but only if the curves intersect. In general, 'Umar Khayyām notes that there may be two admissible solutions (he counts double roots as a single solution), but in Case 13 he has overlooked the possibility that all three solutions may be positive.

Example 1. Construction for Case 2: $x^3 + bx = c$ ("A cube and sides are equal to a number").[79]

[79]See pp. 32–34 (and pp. 46–47 for Example 2) in the book cited above, in note 74. "Sides" (*aḍlā'*), that is, sides of the cube (like "roots" were used for squares in al-Khwārizmī's six equations, see page 57).

Equation $(a, b, c > 0)$:	Solutions:	Curves:
1. $x^3 = c$	$x_1 > 0$; $x_{2,3} \in \mathbb{C}$	P, P
2. $x^3 + bx = c$	$x_1 > 0$; $x_{2,3} \in \mathbb{C}$	C, P
3. $x^3 + c = bx$	$x_{1,2} > 0$ or $\in \mathbb{C}$; $x_3 < 0$	P, H
4. $x^3 = bx + c$	$x_1 > 0$; $x_{2,3} < 0$ or $\in \mathbb{C}$	P, H
5. $x^3 + ax^2 = c$	$x_1 > 0$; $x_{2,3} < 0$ or $\in \mathbb{C}$	P, H
6. $x^3 + c = ax^2$	$x_{1,2} > 0$ or $\in \mathbb{C}$; $x_3 < 0$	P, H
7. $x^3 = ax^2 + c$	$x_1 > 0$; $x_{2,3} \in \mathbb{C}$	P, H
8. $x^3 + ax^2 + bx = c$	$x_1 > 0$; $x_{2,3} < 0$ or $\in \mathbb{C}$	C, H
9. $x^3 + ax^2 + c = bx$	$x_{1,2} > 0$ or $\in \mathbb{C}$; $x_3 < 0$	H, H
10. $x^3 + bx + c = ax^2$	$x_{1,2} > 0$ or $\in \mathbb{C}$; $x_3 < 0$	C, H
11. $x^3 = ax^2 + bx + c$	$x_1 > 0$; $x_{2,3} < 0$ or $\in \mathbb{C}$	H, H
12. $x^3 + ax^2 = bx + c$	$x_1 > 0$; $x_{2,3} < 0$ or $\in \mathbb{C}$	H, H
13. $x^3 + bx = ax^2 + c$	$x_1 > 0$; $x_{2,3} > 0$ or $\in \mathbb{C}$	C, H
14. $x^3 + c = ax^2 + bx$	$x_{1,2} > 0$ or $\in \mathbb{C}$; $x_3 < 0$	H, H

TABLE 1

Take $p = \sqrt{b}$ and $q = \frac{c}{b}$. So the equation becomes $x^3 + p^2x = p^2q$. Draw a (semi)circle of diameter q and a parabola with (double) parameter p and vertex intersecting the circle perpendicularly at one endpoint of the diameter (see Figure 28).[80]

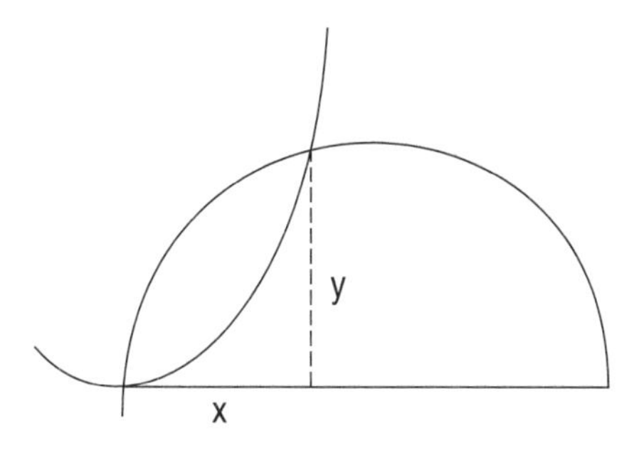

FIGURE 28

By the property of the circle (or by the theorem on the altitude of an inscribed right triangle), we have

$$\frac{q - x}{y} = \frac{y}{x}.$$

[80]In Greece (and therefore in the Middle Ages), the parameter of the parabola was considered to be the segment p, thus the *latus rectum* or twice what we call the "parameter" today.

On the other hand, by the property of the parabola, we also have

$$\frac{y}{x} = \frac{x}{p}.$$

Therefore

$$\frac{q-x}{y} = \frac{x}{p},$$

so

$$pq - px = xy = \frac{x^3}{p}.$$

It follows that

$$x^3 + p^2x = p^2q.$$

Therefore, the segment x corresponding to the intersection is the desired solution. "To this type belong neither a variety of cases nor impossible problems, and it is solved by means of the properties of the circle together with those of the parabola," concludes 'Umar Khayyām. The first part of this statement means that an equation of this type *always* has one real solution, which is *always* positive. See Appendix C.4 for the original text of this example.

Example 2. Construction for Case 8: $x^3 + ax^2 + bx = c$.

This time, take $q = \sqrt{b}$ and $p = \frac{c}{b}$. The equation then becomes $x^3+ax^2+q^2x = pq^2$. Then draw the segment q, and from one of its endpoints draw a perpendicularly in one direction and p perpendicularly in the other; then draw a line parallel to this segment through E, the other endpoint of q (see Figure 29). Next, draw a semi-circle with diameter $a+p$ and an equilateral hyperbola through the endpoint of p with asymptotes being the extension of q and the horizontal line through E.

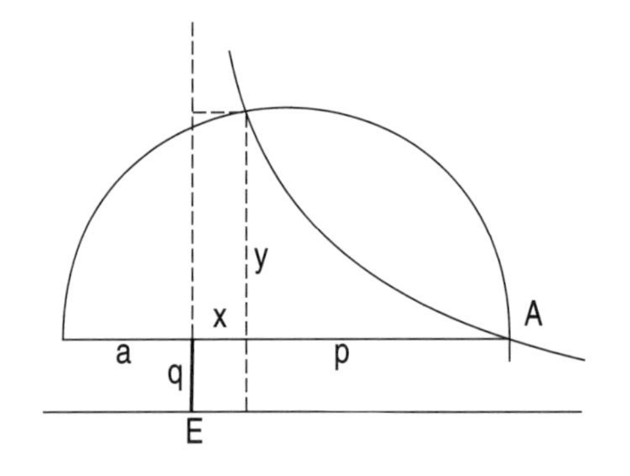

FIGURE 29

By the property of the hyperbola, we have

$$xy = pq.$$

Subtracting qx from both sides yields

$$x(y-q) = q(p-x),$$

so

$$\frac{y-q}{p-x}=\frac{q}{x},$$

and thus

$$\frac{(y-q)^2}{(p-x)^2}=\frac{q^2}{x^2}.$$

Now, by the property of the circle, we have

$$(y-q)^2=(x+a)(p-x).$$

Using this and the previous result, we obtain

$$\frac{x+a}{p-x}=\frac{q^2}{x^2},$$

and thus

$$x^3+ax^2+q^2x=pq^2.$$

Then segment x, corresponding to the intersection, is the desired solution. "To this type belong neither a variety of cases nor impossible problems, and it is solved by means of the properties of the hyperbola together with those of the circle," Khayyām concludes for the same reasons as in the previous example.

Chapter 4

Algebra in Medieval Europe

4.1. Introduction

Early medieval Europe knew almost nothing of the mathematics of antiquity. Greek was no longer the language of culture, and only some extracts of Euclid's *Elements*, an adaptation of Nicomachus's *Introduction to Arithmetic* by Boethius (see page 54), as well as some works on surveying (page 21) existed in Latin translation. Problems on practical geometry and recreational mathematics in the Carolingian collection by Alcuin (page 78) also appear to have had origins in antiquity. These were the only mathematical traces left by antiquity in early medieval Europe. Thus the only possibility of recovering more significant remains lay in transmission from outside.

The first opportunity came about with the Christian reconquest of Spain. It was there, notably in Toledo, that the principal encounter with both Greek science (in Arabic translation) and Muslim science occurred in the twelfth century, at least with what was available in Spain at the time. The most important Greek works (Euclid's *Elements* and, for astronomy, Ptolemy's *Almagest*) were translated; then, al-Khwārizmī's *Arithmetic* and *Algebra* (the latter partially, with the sections on geometry and inheritance algebra omitted); finally, various smaller treatises, as well as writings on plane and spherical trigonometry, optics and practical geometry of ancient or Arab origin. It was also there that the first commentaries and original works were written, although these were, of course, still entirely dependent on works existing in Arabic. The most important of them is the *Liber mahameleth*, of which the name reveals the work's sources while announcing its subject, as *mu'āmalāt* is the Arabic name for mathematics applied to commerce.

Here it should be noted that the Latin translations of mathematical texts were sometimes written by people who were learning the material while translating it, or who pursued the latter activity while neglecting the former. Thus, the quality of the translations often left much to be desired, but, strange though it may seem, this was of secondary importance. First, the translations are completely literal: the word order is Arabic, the expressions are Arabic, and sometimes even the syntax is Arabic. (Incidentally, for today's scholars, a word-by-word translation from Latin back to Arabic is, for that very reason, fairly easy.) But passages where the text is truly incomprehensible are rare. The problem becomes more serious when translators, in their ignorance of the content, choose from among multiple possibilities for the same word precisely the least suitable translation. These various features appear in the (partial) translation of Abū Kāmil's *Algebra*, which dates from the fourteenth century but reproduces the practices in use two centuries earlier. As the preserved copy is in the handwriting of the translator, one can follow his hesitations (through his erasures, alterations, and various attempts at translating the same word) to the point where he starts to waver in his painstaking efforts before abruptly deciding to put an end to them.

For literary texts, such a method of working would be quite inacceptable. But for scientific texts, word-by-word translation was not so bad: by remaining close to the original text, it did not distort its meaning, and serious misinterpretations, arising only in passages of loftier style—introductions, generalities, and various digressions—detracted little from the mathematical substance. In other words, while these translations were far from satisfying any criteria for literary quality, they supplied mathematical tools where there had been none before. And they achieved their purpose, as only a few decades sufficed for Christian Europe to achieve fundamental knowledge of mathematics.

Here might be drawn an analogy between the role of Baghdad in the ninth century and that of Toledo in the twelfth. However, in the latter case, the role was more limited, in both its accomplishments and their influence. Moreover, Toledo was not unique, as translations were being made in several other places as well, both in Spain and abroad. The most notable example is certainly Sicily, where ties with the Muslim East and Byzantium provided indirect access to the science of this Muslim East and works from Ancient Greece in the original language. Nonetheless, Spain, offering direct access to texts, was favored, and Sicilian resources would only be used to full advantage later. Spain's influence was, however, considerably limited by the fact that it offered neither the richness of Baghdad's Greek sources, nor, it seems, the variety of Arabic texts that could have compensated for this shortcoming. In any event, the results of transmission of mathematical knowledge through Spain were uneven, as what survived was primarily geometry and arithmetic. The influence for algebra is much less evident. This is the case first for al-Khwārizmī's *Algebra*, even though it was translated several times, and second, and even more so, for some other translations, such as a work on applications of algebra to elementary geometry and a collection of problems: they were hardly used and gradually even forgotten altogether[81]. Finally,

[81]The former, the *Liber mensurationum*, was published by H. Busard in the *Journal des Savants*, 1968, pp. 65–124 and the latter, the *Liber augmenti et diminutionis*, by G. Libri (note 49), I, pp. 304–71.

the *Liber mahameleth* enjoyed no more enviable a fate, and its influence was negligible.

If any blame were to be assigned for the situation, it should fall on a single person who, having had the opportunity to perfect his mathematical studies in the Muslim East and in the Byzantine Empire at the turn of the thirteenth century, progressed far beyond his contemporaries. Thus the works of the Pisan Leonardo Fibonacci quickly came to replace those known from Spain. Indeed, arriving at just the right time and in the right place, they were destined to do so from the outset. Italy was then blossoming intellectually: it was ready to acquire, and to bring progress to, theoretical mathematics. At the same time, Italy's merchants, trading intensely throughout the Mediterranean, had a pressing need for mathematics applied to commerce; furthermore, the introduction of the new (Indian) numerals and the new arithmetic was greatly facilitated by the growing use of accounting books, which in turn was encouraged by the introduction of paper to replace expensive parchment.

From then on, Italy occupied a dominant position in Europe for commercial arithmetic and algebra, which it maintained throughout the next three centuries. The influence of Fibonacci was lasting, and the works of his successors often include the very same problems, or at least extremely similar ones. In the fourteenth century, these disciples also became serious rivals: their works were simpler and written in Italian, and thus accessible to a broader (and less educated) readership; moreover, being more recent, references to Mediterranean measures and currencies were up-to-date. From the end of the fifteenth century, the printing of treatises, in particular Luca Pacioli's vast *Summa*, which was both theoretical and applied and which incorporated the topics of almost all of Fibonacci's writings, made the earlier works obsolete.[82] In any case, Fibonacci's theoretical knowledge was destined to fall into disuse in the sixteenth century, when methods of solving algebraic equations were extended to the cubic and quartic cases. Thus after superseding the works of the twelfth century, Fibonacci's writings came to be forgotten as well, and they were not published until the nineteenth century[83].

4.2. The *Liber mahameleth*

The fate of the *Liber mahameleth* was even less enviable than that of the works of Fibonacci. There are no more than a handful of references made to it during the Middle Ages, and any attempts to find references from the beginning of modern times until recent years would be in vain. Nonetheless,

[82] *Summa de arithmetica, geometria, proportioni e proportionalità*, Venice 1494 (2nd ed. Toscolano 1523).

[83] *Scritti di Leonardo Pisano, matematico del secolo decimoterzo*, ed. B. Boncompagni (2 vol.), Rome 1857–62. There is now an English (very literal and uncommented) translation by L. Sigler, *Fibonacci's Liber Abaci*, New York 2002.

it was the most extensive and original work on commercial arithmetic and algebra produced during the twelfth century, and its presumed author, John of Seville (Johannes Hispalensis), was one of the most illustrious translators in Toledo in the 1140s. The *Liber mahameleth* is also our best source on the state of arithmetic, algebra, and their applications in Muslim Spain at the beginning of the twelfth century, the documents hitherto studied in Arabic and Latin translation being too scarce to allow us to precisely reconstruct the state of knowledge.[84]

The *Liber mahameleth* is a curious blend, and the fact that it received so little attention is perhaps also because it failed to truly satisfy anyone: the mathematician, who would certainly enjoy the proofs based on Euclid, would dislike the interminable succession of applied problems on single subjects or solutions giving simple calculations in minute detail; on the other hand, a merchant knowing only the rudiments of mathematics would be able to follow the calculations, but he would be lost when facing Euclidean-like demonstrations and calculations involving expressions containing irrational quantities.

Indeed, we have seen that in his *Algebra*, Abū Kāmil calculated with complicated irrational quantities and used geometric figures to illustrate the solution of quadratic equations and the treatment of problems. After him, in the Muslim East, algebra gradually gained its independence from geometry as algebraic reasoning came to be considered valid in itself. Meanwhile, the Muslim West, which for both geographic and political reasons knew little of Eastern developments since the middle of the tenth century, took no part in this evolution. Instead, it remained faithful to the spirit of Abū Kāmil's book, and thus the *Liber mahameleth*, which explicitly refers to Abū Kāmil's *Algebra* in several instances and is implicitly influenced by it in numerous others, did so too. That is why, besides (a few) computations involving awkward irrational quantities, we encounter geometric illustrations throughout the text, notably more than in Abū Kāmil's *Algebra*. This goes so far that each treatment of a new type of problem often contains three separate parts: a formula, given without explanation, to be applied directly; a geometric solution showing that this formula can be established via Euclidean geometry; and, finally, an algebraic solution (we have seen an instance of the latter two parts in Abū Kāmil). As for the form, the algebra in the *Liber mahameleth* already had the characteristics it would retain throughout the Middle Ages. Following the model of texts in Arabic, no symbolism is used, though particular names are used to distinguish the unknown and its powers: *res* for x, *census* for x^2, *cubus* for x^3 (and, although this does not occur in the *Liber mahameleth*, combinations of *census* and *cubus* for the higher powers). These terms are simply the Latin translations of their Arabic counterparts.

[84]This may change: mathematical texts from Islamic Spain and North Africa are presently being studied intensively by A. Djebbar and his students.

The *Liber mahameleth* also writes numbers out as words, but this habit was not to last.[85]

Example 1. *Liber mahameleth*, Problem B.185 from Chapter B–IX on boiling must (*capitulum de coquendo musto*)[86] [Appendix B.8]

Among the various groups of applied problems in the *Liber mahameleth*, we find one subject which, while departing from the usual questions of sales, expenditures, and the hiring of workers, is still an application problem and not one of the recreational group that closes the work. This is the subject of boiling must. This kind of problem is not known from any other medieval European source, apart from a few Italian and French texts which took it over from the *Liber mahameleth* and often made them problems on boiling *wine*, thus completely distorting the meaning of the original problem. Indeed, boiling must corresponds to a specific Islamic tradition since it has its origin in the prohibition of alcoholic beverages. A beverage obtained by boiling must will be considered as legal (*ḥalāl*) when reduced to a certain fraction of the original quantity, a third according to some experts in law and holy tradition. For that purpose, a flat-bottomed cylindrical vat was used, divided into three equal parts by two engraved circles and smaller marks for finer graduation. Now it may happen, during the boiling process, that part of the must overflows. We are then to pursue the reduction process until a new remainder is obtained which will be neither the original third, since we now have less liquid than before, nor one third of the remaining must, since part of the reduction has already been completed.

Let the initial quantity of must be q, which is to be reduced to r_1 through the evaporation of the quantity $q - r_1$. Now during this operation, once the liquid has already been reduced by d, v happens to boil over. In the same way as the quantity $q - d$ would have to be boiled down to r_1, the remaining quantity $q - d - v$ must be reduced to r_2; that is, the ratios must be equal. Thus we must have

$$(*) \qquad \frac{q-d}{r_1} = \frac{q-d-v}{r_2}.$$

Fourteen problems concern this boiling of must. In the tenth, one must calculate the initial quantity q of must given that $r_1 = \frac{1}{3}q$, $d = 2$, $v = 2$, and $r_2 = 2 + \frac{1}{2}$.

(1) As we mentioned above, the treatment begins with the computation of the solution by the appropriate, but still unexplained, formula. In modern mathematical language, this formula is (writing, generally,

[85] The *Liber mahameleth* will be published in the series *Sources in the History of Mathematics and Physical Sciences* (Springer-Verlag).

[86] Book A of the *Liber mahameleth* contains the mathematical theory, Book B the applications.

$r_1 = \frac{1}{k}q)$

$$q = \frac{1}{2}(kr_2 + d + v) + \sqrt{\left[\frac{1}{2}(kr_2 + d + v)\right]^2 - kr_2 d}.$$

We thus find that $q = 10$.

(2) For the medieval mathematician, and not only the one intrigued by the preparation of this kind of beverage, having just a formula would not have sufficed. Therefore, the next step is to prove it geometrically. Thus let $\mathrm{AB} = q$, $\mathrm{AD} = d = 2$, $\mathrm{DG} = v = 2$, $\mathrm{KB} = r_2 = 2 + \frac{1}{2}$, and $\mathrm{HB} = 3r_2 = 7 + \frac{1}{2}$ (see Figure 30).[87] By (*),

$$\frac{q-d}{3r_1} = \frac{q-d}{q} = \frac{q-d-v}{3r_2},$$

so we have

$$\frac{\mathrm{DB}}{\mathrm{AB}} = \frac{\mathrm{GB}}{\mathrm{HB}} < 1.$$

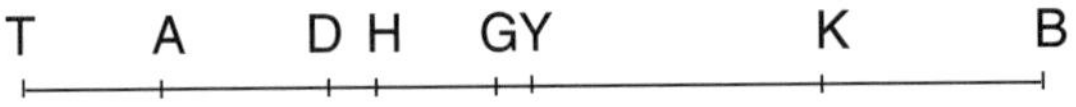

T A D H GY K B

FIGURE 30

From this equality of ratios, it follows that

$$\frac{\mathrm{AB} - \mathrm{DB}}{\mathrm{AB}} = \frac{\mathrm{AD}}{\mathrm{AB}} = \frac{\mathrm{HB} - \mathrm{GB}}{\mathrm{HB}} = \frac{\mathrm{HG}}{\mathrm{HB}},$$

that is,

$$\frac{q-(q-d)}{q} = \frac{d}{q} = \frac{3r_2 - (q-d-v)}{3r_2},$$

and thus that

$$\mathrm{AB} \cdot \mathrm{HG} = \mathrm{AD} \cdot \mathrm{HB} = 15,$$

that is,

$$q \cdot [3r_2 - (q-d-v)] = d \cdot 3r_2 = 15.$$

We may observe that not only do we know the product of q and $3r_2 - (q-d-v)$ but we also know their sum, since it is $3r_2 + d + v = 11 + \frac{1}{2}$. So, to represent this sum, let us add to $q = \mathrm{AB}$ the quantity $3r_2 - (q-d-v) = \mathrm{HG}$, that is, put $\mathrm{TA} = \mathrm{HG}$, thus forming $\mathrm{TA} + \mathrm{AB} = \mathrm{TB} = 11 + \frac{1}{2}$. What we now have is a nice mixture of ancient mathematical techniques: the Mesopotamian algebra of identities and Euclidean theoretical geometry. We are to solve

$$\begin{cases} \mathrm{AB} + \mathrm{TA} = 11 + \frac{1}{2} \\ \mathrm{AB} \cdot \mathrm{TA} = 15. \end{cases}$$

[87] We will slightly simplify the proof given in the original text by representing all the values on the same line segment.

We can therefore apply the identity

$$\left(\frac{\mathrm{AB}-\mathrm{TA}}{2}\right)^2=\left(\frac{\mathrm{AB}+\mathrm{TA}}{2}\right)^2-\mathrm{AB}\cdot\mathrm{TA}.$$

Then we know the term on the left side, whence

$$\mathrm{AB}=\frac{\mathrm{AB}+\mathrm{TA}}{2}+\frac{\mathrm{AB}-\mathrm{TA}}{2}.$$

This is what our author does in his geometrical demonstration. With Y as the midpoint of TB, we have a segment of straight line (TB) divided once into two equal parts (at Y) and once into two unequal parts (at A). Thus we can apply Proposition II.5 of the *Elements* (page 65) to obtain

$$\mathrm{TA}\cdot\mathrm{AB}+\mathrm{AY}^2=\mathrm{TY}^2.$$

As $\mathrm{TA}\cdot\mathrm{AB}=15$ and $\mathrm{TY}^2=\left(\frac{\mathrm{TB}}{2}\right)^2=33+\frac{1}{16}$, we find $\mathrm{AY}=4+\frac{1}{4}$, and, since $\mathrm{YB}=5+\frac{3}{4}$,

$$\mathrm{AB}=\mathrm{AY}+\mathrm{YB}=10.$$

The author has thus proved the formula used at the beginning. For

$$\mathrm{AY}=\sqrt{\mathrm{TY}^2-\mathrm{TA}\cdot\mathrm{AB}}=\sqrt{\left(\frac{1}{2}\mathrm{TB}\right)^2-\mathrm{HG}\cdot\mathrm{AB}}$$

$$=\sqrt{\left(\frac{1}{2}\mathrm{TB}\right)^2-\mathrm{HB}\cdot\mathrm{AD}}=\sqrt{\left[\frac{1}{2}\left(3r_2+d+v\right)\right]^2-3r_2d},$$

while

$$\mathrm{YB}=\frac{1}{2}\mathrm{TB}=\frac{1}{2}\left(\mathrm{HB}+\mathrm{TH}\right)=\frac{1}{2}\left(\mathrm{HB}+\mathrm{AG}\right)$$

$$=\frac{1}{2}\left(\mathrm{HB}+\mathrm{AD}+\mathrm{DG}\right)=\frac{1}{2}\left(3r_2+d+v\right).$$

(3) Finally, for the pure algebraic treatment, set $q=x$ (*res*). From the fundamental proportion (*) for this class of problem (page 97) and the givens of this particular problem, we have

$$\frac{x-2}{\frac{1}{3}x}=\frac{x-4}{2+\frac{1}{2}},$$

which yields

$$x^2+15=\frac{23}{2}x,$$

the (positive and admissible) solution of which is $x=10$.

Example 2. *Liber mahameleth*, Problem B.315 from Chapter B–XVIII on the exchange of morabitini (*capitulum de cambio morabitinorum*) [Appendix B.9]

The *morabitinus*, which became the Spanish maravedi, was a gold piece minted in Christian Spain that replaced the equivalent in use in Muslim Spain under the dynasty of the Almoravides (Arabic *al-Murābiṭūn*, whence the coin's name). Since, however, the coinage at that time was local, the relative values of morabitini and other pieces were local as well, but the wide circulation of coins made it likely that an exchange would involve different kinds of morabitini as well as smaller coins having the same name but different values. Such is the subject of the exchange problems in the *Liber mahameleth*. In the present problem, one gold piece is traded for two different types of *nummi*, and 30 of one and 10 of the other are each worth this gold piece. As a result of the exchange, we are told, the number of the first coins exceeds the number of the second by 20.

In our symbols, this is equivalent to solving the linear system

$$\begin{cases} a_1x_1 - a_2x_2 = b \\ x_1 + x_2 = 1, \end{cases}$$

with $a_1 = 30$, $a_2 = 10$, and $b = 20$. We wish to find the quantities a_1x_1 and a_2x_2 of each coin.

(1) As in the problem above, the answer is first calculated using a formula given without explanation:

$$x_2 = \frac{a_1 - b}{a_1 + a_2} \qquad (\text{and } x_1 = 1 - x_2),$$

whence a_1x_1 and a_2x_2.

(2) To establish this relation geometrically, the author takes (see Figure 31) DB $= x_1$, AD $= x_2$, with AB $= 1$. Now draw the segments DK $= a_1$ and DG $= a_2$ perpendicularly to AB; then complete the rectangle. We thus have that KB $= a_1x_1$ and AG $= a_2x_2$. Then form a rectangle DH on BD with area equal to that of AG. (*Elements* I.43 proves that in a parallelogram, the two "complementary" parallelograms formed by two line segments parallel to the sides, and intersecting on the diagonal, are equal in area.)[88] We thus find that KH $=$ KB $-$ DH $=$ KB $-$ AG $= a_1x_1 - a_2x_2 = b$. On the other hand, IB $=$ AB $\cdot$ DK $= a_1$. This yields the formula for x_2, since $a_1 - b =$ IB $-$ KH $=$ DI $+$ DH $=$ DI $+$ AG $=$ IG $= (a_1 + a_2)x_2$. From the figure we likewise infer that DB $\cdot$ GK $= x_1(a_1 + a_2) =$ KH $+$ HG $=$ KH $+$ BT $= b + a_2$.

[88]Such is the construction for obtaining a rectangular area equal to another one, when the base of the first is imposed. In the *Liber mahameleth*, the author just reproduces the rectangle AG within the rectangle KB (see Appendix B.9)—thus departing from his usual rigor.

(3) Finally, the algebraic treatment takes our x_2 to be the unknown x and substitutes $1-x$ for x_1 in the first equation, whence

$$30(1-x)-10x=30-40x=20.$$

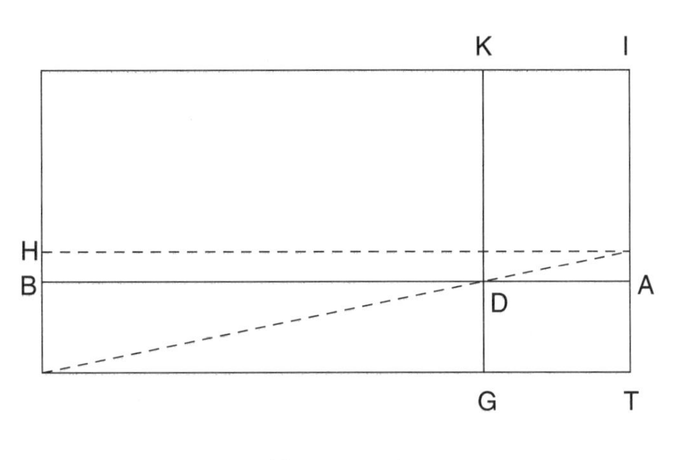

FIGURE 31

4.3. Leonardo Fibonacci

Introduction

Leonardo of Pisa, or Leonardo Fibonacci—the introduction of patronymic names was recent at the time—was born around 1170 in Pisa, where he died soon after 1240. He himself tells us about the origin of his vocation in the introduction to his major work on arithmetic and algebra, the *Liber abaci*. His father worked for the Pisan merchants at the customs of Bougie (Algeria), where they frequently came for trade. Aware of the future advantages this would give his son, his father had him, "still a child" (*in pueritia*), come for some time in order to be instructed in the science of arithmetic. Leonardo was, in his own words, "trained by an admirable teaching in the science using the nine Indian signs" (*ex mirabile magisterio in artem per novem figuras Indorum introductus*)[89]. This first encounter with mathematics must have been some kind of revelation. For, upon reaching adulthood, he was forever seeking to deepen his knowledge, both through his own studies and by direct contact with scholars met on his travels, when business took him to Egypt, Syria, and Byzantium. He also cultivated more local relations. We know from other works, for example, that during the 1220s he was in close contact with scholars at Frederick II's court in Sicily.

His *Liber abaci* is a vast work of fifteen chapters that explains arithmetic and algebra, as well as how to solve numerous problems of applications to

[89]In the Middle Ages, one usually referred to the "nine signs": zero is not included.

trade, and also problems of the recreational kind, describing irrealistic situations. This reminds us of the treatise we have just seen. Indeed, what Leonardo Fibonacci calls *abacus* is essentially the same as what John of Seville calls *mahameleth*—for, although the primary meaning of *abacus* was the calculating board, it came to mean, by extension, calculation itself. Fibonacci, however, presents a greater variety of problems, particularly appropriate for trade, in a treatise coming, as said before (page 95), at the right time and in the right place. The importance of the *Liber abaci* is best illustrated by the fact that the Italian treatises on commercial mathematics over the next three centuries usually bear the name *trattati d'abaco* and that a school teaching this subject was a *bottega d'abaco*.

Today Leonardo Fibonacci is primarily known for the sequence that bears his name, which appears in one of the problems in his *Liber abaci*—although none of the sequence's properties are mentioned. In fact, the problem is more recreational in nature, for it does not reflect any real situation. It asks how many pairs of rabbits will result after one year from a single pair of rabbits, shut up in an enclosure, under the following hypotheses:

- The interval between two generations, the gestation time, and the time to adulthood are invariably one month.
- According to behavior that has long been observed, the rabbits mate frequently, in particular immediately after giving birth.
- Each litter is a pair of rabbits, one male and one female.
- No rabbit dies or escapes during the year.

Let us follow the development from a single pair P_1 starting at birth. As there is an interval of one month between birth and adulthood, and then another month before it gives birth to a new pair, the offspring P_{11} of P_1 will be born two months after the birth of P_1. One month later, P_1 gives birth to a new pair P_{12}, as well as in every month thereafter. The first pair born to P_{11} comes into the world in the same month that P_{13} is born. We then see that the number u_m of pairs in month m equals the number u_{m-1} of pairs existing in month $m-1$ (as there are no deaths or escapes), plus the number of pairs produced by the u_{m-2} pairs present in month $m-2$. This yields the recurrence relation

$$u_m = u_{m-1} + u_{m-2}.$$

If we start this sequence with the birth of P_1, the terms are

$$1, 1, 2, 3, 5, 8, 13, 21, \ldots$$

Fibonacci, who begins the year with the first birth to the pair P_1, finds that the number of pairs at the end of the year is 377 ($= u_{14}$).[90]

[90] *Scritti* (note 83), I, pp. 283–84. English translation, pp. 404–5.

Leonardo Fibonacci's Linear Systems

Fibonacci's fame during the Middle Ages had in no way to do with this rabbit problem, as it does today (at least among mathematicians). His strongest mathematical influence lay in what seems to have been his favorite topic, solving determinate and indeterminate systems of linear equations in which the unknowns represent concrete quantities (most often sums of money). As well as forming a considerable portion of his *Liber abaci*, these systems appear in two minor works: the *Letter to Theodore* (*Epistola ad Theodorum*) and the *Flower* (*Flos*). It would seem that Fibonacci's intensive research on the subject is the result of his Byzantine sojourns. In any case, the solution of such a problem in the way we shall see is explicitly attributed by him to a "most able" Byzantine scholar named *Muscus* (clearly the Greek name Μόσχος) of whom we know nothing today.[91]

Fibonacci classifies these systems, and in many cases establishes a general solution formula for each type (although without using any sort of algebraic symbolism). His method of obtaining this formula—completing the equations so that each incorporates the sum of all the unknowns, then combining the new equations with the original ones—was not new, as such a procedure had already been in use in antiquity with an application of the rule of Thymaridas (pages 26–27). However, in a few (rare) cases, it is clear that Fibonacci has used his formulas to deliberately choose givens that lead to a negative value for one or more of the unknowns. These then are not simply subtracted values, the use of which in calculations is as old as algebra, but true negative quantities, left as such since no further operation is acting on them. This attitude is new. However, as we will see in a few examples, Fibonacci fails to distinguish himself from his predecessors, as he too rejects any negative solution as such. But, and this again is new, he attempts to interpret the negative solution, in the framework of the concrete situation represented by the problem, as a positive quantity which will then be subtracted in the proposed equations. This distinction is not as trivial as it may first seem: it showed that a negative result can be meaningful, that is, represent a real situation, which later opened the door to the acceptance of negative numbers in their own right. First, though, we will consider the general form of his main systems in modern notation and the method of solving them.[92]

[91] *Questio nobis proposita a peritissimo magistro musco constantinopolitano in constantinopoli*, as the text in *Scritti*, I, p. 249 reads. The English translation, p. 362, speaks of "a most learned master of a Constantinople mosque", which is absurd.

[92] These systems are studied by K. Vogel, "Zur Geschichte der linearen Gleichungen mit mehreren Unbekannten," *Deutsche Mathematik*, 5 (1940), pp. 217–40 and (particularly treating negative solutions) in J. Sesiano, "The appearance of negative solutions in mediaeval mathematics," *Archive for history of exact sciences*, 32 (1985), pp. 105–50. For references to the presence of such problems, see the most recent edition of Tropfke's *Geschichte der Elementarmathematik* (note 69).

(1) Exchanging Sums of Money

These problems were already known in antiquity, at least in a simpler form: they are seen in the epigrams of the *Greek Anthology* (see page 25), in Book I of the *Arithmetica* (among well-known school problems, and in problems dealing with pure numbers), as well as in Alcuin's collection (page 78). This type of problem is also found in Chinese and Indian mathematics, as well as in treatises in Arabic[93]. But it was with Leonardo Fibonacci that their development reached its height.

The general problem considers a group of n people each having a certain sum, say x_i for the ith person. One by one, a group of $t+1$ people with t fixed $(0 \leq t \leq n-2)$, taken cyclically starting with the jth person $(1 \leq j \leq n)$, borrows a given sum z_j from the $n-(t+1)$ other people. Given for each exchange the linear relation linking the amount held by the first group with the amount remaining to the second group (with multiplier m_j and additional difference y_j), one is asked to determine the initial assets of each person.

Thus we have n equations

$$\sum_{l=j}^{j+t} x_l + z_j = m_j \left(\sum_{i \neq j,\ldots,j+t} x_i - z_j \right) + y_j, \qquad \text{for } j = 1, \ldots, n,$$

where m_j, y_j, and z_j are given. Rewrite this as

$$\sum_{l=j}^{j+t} x_l + z_j - y_j = m_j \left(\sum_{i \neq j,\ldots,j+t} x_i - z_j \right)$$

and add $\sum_{i \neq j,\ldots,j+t} x_i - z_j$ to each side. Then, by setting $S = \sum_{i=1}^{n} x_i$, we have

$$\text{(1)} \qquad S - y_j = (m_j + 1) \left(\sum_{i \neq j,\ldots,j+t} x_i - z_j \right).$$

This yields

$$(*) \qquad \sum_{i \neq j,\ldots,j+t} x_i = \frac{S - y_j}{m_j + 1} + z_j.$$

Now add these n equations. As each sum on the left side lacks $t+1$ consecutive terms, thus containing $n-(t+1)$ consecutive unknowns x_i, and the indices are taken cyclically, each x_i appears in the result $n-(t+1)$ times altogether; and so, therefore, does S. Hence

$$(**) \qquad S = \frac{1}{n-(t+1)} \left[\sum_{i=1}^{n} \frac{S - y_i}{m_i + 1} + \sum_{i=1}^{n} z_i \right],$$

[93] Tropfke, *op. cit.* (note 69), pp. 610–11.

and solving for S yields

$$S = \frac{\sum z_i - \sum \frac{y_i}{m_i+1}}{[n-(t+1)] - \sum \frac{1}{m_i+1}}. \tag{2}$$

On the other hand, as

$$\sum_{l=j}^{j+t} x_l = S - \sum_{i \neq j,\ldots,j+t} x_i,$$

we have by (*) and (**) that

$$\sum_{l=j}^{j+t} x_l = \frac{1}{n-(t+1)} \sum_{i=1}^{n} \frac{S-y_i}{m_i+1} - \frac{S-y_j}{m_j+1} + \frac{1}{n-(t+1)} \sum_{i=1}^{n} z_i - z_j. \tag{3}$$

We are thus led to solving n equations of the form

$$\sum_{l=j}^{j+t} x_l = r_j,$$

where the r_j are known since y_i, m_i, z_i are given and S has been found by (2). From the rewritten initial equation, we see that for the exchange to be possible we must have

$$\sum_{i \neq j,\ldots,j+t} x_i - z_j \geq 0$$

for all j, which means, by (1), that $S \geq y_j$ for all j.

Example 1. *Liber abaci*, simple case "on two men" (*de duobus hominibus*) from Chapter XII.3 (*Scritti*, I, 192–98)

Consider the following case of two people exchanging a certain sum:

$$\begin{cases} x_1 + 7 = 5(x_2 - 7) + y_1 \\ x_2 + 5 = 7(x_1 - 5) + y_2. \end{cases}$$

We introduce the auxiliary quantities

$$\begin{aligned} S &= x_1 + x_2 \\ s_1 &= x_1 + x_2 - y_1 = S - y_1 \\ s_2 &= x_1 + x_2 - y_2 = S - y_2. \end{aligned}$$

This is what Fibonacci does, except that, as he uses no symbols, he refers to each quantity by a specific name (*maior summa*, *mediana summa*, *minor summa*). Next, in the proposed system

$$\begin{aligned} x_1 + 7 - y_1 &= 5(x_2 - 7) \\ x_2 + 5 - y_2 &= 7(x_1 - 5), \end{aligned}$$

we add $x_2 - 7$ and $x_1 - 5$ to both sides of the two equations, respectively. Using the symbols introduced above, we then obtain, as in (1) before,

$$s_1 = 6(x_2 - 7)$$
$$s_2 = 8(x_1 - 5),$$

so, as in (*),

$$x_1 = \frac{1}{8}s_2 + 5$$
$$x_2 = \frac{1}{6}s_1 + 7.$$

Thus, as in (**),

$$\begin{aligned} x_1 + x_2 &= S \\ &= \frac{1}{6}s_1 + \frac{1}{8}s_2 + 12 \\ &= \frac{1}{6}(S - y_1) + \frac{1}{8}(S - y_2) + 12. \end{aligned}$$

From this we find that, as in (2),

$$S = \frac{24}{17}\left[12 - \left(\frac{1}{6}y_1 + \frac{1}{8}y_2\right)\right].$$

As we know y_1 and y_2, we know S, and thus s_1 and s_2, and finally x_1 and x_2 as well. However, as remarked above, we do need the two quantities

$$s_1 = S - y_1 = \frac{24}{17}\left[12 - \left(\frac{21}{24}y_1 + \frac{1}{8}y_2\right)\right] = \frac{3}{17}\left[96 - (7y_1 + y_2)\right]$$

and

$$s_2 = S - y_2 = \frac{24}{17}\left[12 - \left(\frac{1}{6}y_1 + \frac{20}{24}y_2\right)\right] = \frac{4}{17}\left[72 - (y_1 + 5y_2)\right]$$

to be non-negative.

This way of solving the problem is Fibonacci's, except that the use of modern notation makes is considerably shorter; we thus see that Fibonacci follows the general solution as described before. It is true, though, that the principle dates back to antiquity.

Fibonacci distinguishes various cases for this problem, depending on the (integral) values given for the differences y_1 and y_2.

- The differences are equal: $y_1 = y_2$. We then see that the quantities in brackets are positive when $y_1 = y_2 = y < 12$. Fibonacci solves the two extreme cases. He begins with the case $y = 1$[94] and then turns to the case $y = 12$. Our attempts have shown, he says, that such problems are not solvable for $y > 11$ (*ab undecim vero superius*

[94]He solves the case $y_1 = y_2 = 0$ just before this (*Scritti*, I, pp. 190–91). As he states in opening the problem, Fibonacci learned of it in Byzantium (*questio* (...) *nobis apud Constantinopoli a quodam magistro proposita*).

ipsas (questiones) insolubiles esse probabimus). He shows this as follows. For $y = 12$, the equations become

$$\begin{cases} x_1 - 5 = 5\,(x_2 - 7) \\ x_2 - 7 = 7\,(x_1 - 5)\,. \end{cases}$$

From what we saw above,

$$\frac{1}{8}s_2 = x_1 - 5$$
$$\frac{1}{6}s_1 = x_2 - 7,$$

and so, by the first equation,

$$\frac{1}{8}s_2 = \frac{5}{6}s_1,$$

while, since $y_1 = y_2$,

$$s_1 = s_2 = s.$$

Taking the value $y = 12$ thus leads to something which is "impossible" (or "unsuitable": he uses both *impossibile* and *inconveniens*), and all the more so (*multum inconvenientius*) when y is greater than 12. Note, however, that when $y = 12$, we have $s = 0$, a situation that is not impossible because it simply means that the first man had 5, the second 7, and each loans his entire assets to the other. But Fibonacci disregards this possibility.

- The differences are additive: $y_1,\ y_2 > 0$. We see that the signs of the quantities between brackets depend on the choice of the y_i. Fibonacci mentions that such cases cannot always be solved (*notandum quia quedam ex similibus questionibus sunt insolubiles*), and then considers one case that is solvable ($y_1 = 1$, $y_2 = 2$) and one that is not ($y_1 = 1$, $y_2 = 15$, which leads to $x_1 = 4 + \frac{15}{17}$, "which is unsuitable since it is less than the 5 that the second man requests from the first" (*quod est inconveniens cum sint minus de* 5 *quos petit secundus ipsi primo homini*).
- The differences are subtractive (we deliberately avoid the word "negative"): $y_1,\ y_2 < 0$. This class, says Fibonacci, is solvable without restriction (*hee autem questiones in infinitum tendunt solubiles*). He considers the example of $y_1 = -1$, $y_2 = -3$: "Suppose that the first requests 7 from the second and (then) has five times as much as he and 1 less, (and) that the second requests 5 from the first and (then) has seven times as much as he and 3 less."
- The differences y_1 and y_2 have different signs. The existence of a solution can be inferred from what has been seen above, writes Fibonacci (*ex iis autem que dicta sunt satis potest perpendi si proponatur ut uni ipsorum super suam multiplicitatem superet, alteri vero minuat*). He considers the solvable case $y_1 = 6$, $y_2 = -8$.

He concludes that there are among such problems infinitely many with no solution (*sunt ex similibus questionibus infinite que solvi non possunt*). This is the first time in the *Liber abaci* that the question of negative solutions arises. We see that such solutions (as well as solutions equal to zero) are just as much a reason for rejecting a problem as if the equations of the system were inconsistent.

Example 2. *Liber abaci*, problem "on four men" (*de quattuor hominibus*) from Chapter XII.3 (*Scritti*, I, 201) [Appendix B.10]

In several problems, the differences y_i are given to be zero. Here is one such example.

$$\begin{cases} x_1 + x_2 + 7 = 3\,(x_3 + x_4 - 7) \\ x_2 + x_3 + 8 = 4\,(x_4 + x_1 - 8) \\ x_3 + x_4 + 9 = 5\,(x_1 + x_2 - 9) \\ x_4 + x_1 + 11 = 6\,(x_2 + x_3 - 11)\,. \end{cases}$$

Once again, Fibonacci transforms this system by completing the equations to obtain in each the sum S of the unknowns. This yields the system (of which the general form is our relation (*), page 104, with $y_j = 0$)

$$\begin{cases} x_3 + x_4 = \frac{1}{4}S + 7 \\ x_4 + x_1 = \frac{1}{5}S + 8 \\ x_1 + x_2 = \frac{1}{6}S + 9 \\ x_2 + x_3 = \frac{1}{7}S + 11. \end{cases}$$

From the first and third equations, he deduces that

$$S = 16 + \frac{5}{12}S,$$

and thus

$$S = 27 + \frac{3}{7}.$$

Now the other two yield

$$S = 19 + \frac{12}{35}S,$$

and thus

$$S = 28 + \frac{21}{23}.$$

This result is unsuitable (*inconveniens*), remarks Fibonacci, since we first found that the sum is something else. Hence the problem is unsolvable (*insolubilis*). On the other hand, he continues, the following case is solvable (*solubilis*):

$$\begin{cases} x_1 + x_2 + 100 = 3\,(x_3 + x_4 - 100) \\ x_2 + x_3 + 106 = 4\,(x_4 + x_1 - 106) \\ x_3 + x_4 + 145 = 5\,(x_1 + x_2 - 145) \\ x_4 + x_1 + 170 = 6\,(x_2 + x_3 - 170)\,. \end{cases}$$

He transforms this as usual to obtain

$$\begin{cases} x_3 + x_4 = \frac{1}{4}S + 100 \\ x_4 + x_1 = \frac{1}{5}S + 106 \\ x_1 + x_2 = \frac{1}{6}S + 145 \\ x_2 + x_3 = \frac{1}{7}S + 170. \end{cases}$$

By addition, he then determines that $S = 420$, and it remains to solve the system

$$\begin{cases} x_3 + x_4 = 205 \\ x_4 + x_1 = 190 \\ x_1 + x_2 = 215 \\ x_2 + x_3 = 230, \end{cases}$$

the equations of which are no longer inconsistent. As this system is indeterminate, we can split 215 arbitrarily (*ad libitum*) into two parts, x_1 and x_2, and then calculate the other values. Using this method, Fibonacci obtains the solution $x_1 = 100$, $x_2 = 115$, $x_3 = 115$, $x_4 = 90$.

We chose this example because it again illustrates one aspect of Fibonacci's algebra, which is to examine in which situations a proposed system may, according to the choice of constants, lead to an admissible solution. It would certainly be an exaggeration to say that Fibonacci realized that the equations are linearly dependent when $n - (t+1)$ is a divisor of n (because in this case $n - (t+1)$ sums of equations must produce the same result S and $n - (t+2)$ unknowns can be chosen). But it was clear to him that such a situation could arise when the number of equations forming this type of system was even (the number of his equations is always less than 9).[95]

(2) The Discovery of a Purse

Another type of system frequently encountered in the *Liber abaci* consists of those involving the discovery of a purse. A group of n people, each having a certain unknown sum of money, finds n purses. Each group of $t+1$ people, with t fixed ($0 \leq t \leq n-2$), taken cyclically starting from the jth person ($1 \leq j \leq n$), will have, between their own money and the money from the jth purse, a given multiple of what the others possess. (A simplified case is to restrict the problem to a single purse.) Letting x_l be the assets of the lth person, y_j the purse found by the group starting with the jth person, and m_j the given multiple, we have

$$\sum_{l=j}^{j+t} x_l + y_j = m_j \sum_{i \neq j,\ldots,j+t} x_i \qquad j = 1, \ldots, n,$$

[95] *Si homines pares fuerint possunt quandoque insolubiliter proponi* (*Scritti*, I, p. 286). *Si homines fuerint pares, et duo vel plures per ordinem petant reliquis, erunt questiones eorum quandoque solubiles quandoque non* (*ibid.*, p. 251).

or, in the simplified case,

$$\sum_{l=j}^{j+t} x_l + y = m_j \sum_{i \neq j,\ldots,j+t} x_i \qquad j = 1, \ldots, n.$$

The general method for solving these systems is exactly the same as in the case of exchanging sums of money. But we can obtain the solution formula by setting the z's equal to zero and changing the sign of the y's in formula (3) on page 105, thus finding, respectively,

$$\sum_{l=j}^{j+t} x_l = \frac{1}{n-(t+1)} \sum_{i=1}^{n} \frac{S+y_i}{m_i+1} - \frac{S+y_j}{m_j+1}$$

and

$$\sum_{l=j}^{j+t} x_l = (S+y) \left[\frac{1}{n-(t+1)} \sum_{i=1}^{n} \frac{1}{m_i+1} - \frac{1}{m_j+1} \right].$$

The form of such problems has been known since antiquity: Diophantus considers two simple examples in Book I of his *Arithmetica*, with $n = 3$ (I.18) and $n = 4$ (I.19). Later, such problems can be found in India, Muslim countries, and Byzantium[96].

Example 3. *Liber abaci*, last problem "on finding purses" (*de inventione bursarum*) from Chapter XII.4 (*Scritti*, I, pp. 227–28) [Appendix B.11]

On five men and one purse.

Let us suppose that the first and the second have with the purse twice as much as the remaining three men; the second and the third, three times; the third and the fourth, four times; the fourth and the fifth, five times; that the fifth and the first similarly have six times as much as the remaining three men.

From these stipulations, it appears that the third, the fourth, and the fifth man have $\frac{1}{3}$ *of the sum of deniers of the five men and the purse; that likewise the fourth, the fifth, and the first have* $\frac{1}{4}$*; that the fifth, the first, and the second have* $\frac{1}{5}$*; that the first, the second, and the third have* $\frac{1}{6}$*; that the second, the third, and the fourth have* $\frac{1}{7}$*.*

Set their sum with the purse to be 420*—this number being integrally divisible by the parts mentioned above—and take respectively* $\frac{1}{7}$, $\frac{1}{6}$, $\frac{1}{5}$, $\frac{1}{4}$, *and* $\frac{1}{3}$ *of it.*[97] *You will (thus) have* 140 *for the deniers of the third, the fourth, and the fifth; likewise,* 105 *for the deniers of the fourth, the fifth, and the first;* 84 *for the fifth, the first, and the second;* 70 *for the first, the second, and the third; similarly, the deniers of the second, the third, and the fourth will be* 60. *These five numbers, when added, make* 459 *for three times the deniers of the five men, as each has been counted three times in the stated numbers.*

[96] Tropfke, *op. cit.* (note 69), p. 607.
[97] To be read from right to left.

So take $\frac{1}{3}$ *of* 459; *this yields, since a whole number results,* 153 *for the sum of the deniers of the (five) men. Subtract it from* 420; 267 *remains for the deniers of the purse. Then add the deniers of the first, the second, and the third to the deniers of the fourth, the fifth, and the first, that is,* 70 *to* 105. *The result is* 175, *and this is what they have together, the first counted twice. Thus, subtract* 153, *their sum, from* 175; 22 *remains, and this is what the first possesses. Add this and the deniers of the third, the fourth, and the fifth; it makes* 162, *and this is what the first, the third, the fourth, and the fifth possess. But the five together have only* 153. *Thus this problem is unsolvable, unless we put that the second man has a debt of* 9, *which is (the amount) from* 153 *to* 162. *So add together* 22 *deniers and the debt of the second, that is, subtract* 9 *from* 22, 13 *remains; subtract it from* 70, 57 *remains, and this is what the third possesses. Subtract* 9, *the debt of the second, from it,* 48 *remains; subtract it from the deniers of the second, the third, and the fourth, thus from* 60, 12 *remains, and this is what the fourth possesses. Add it to* 57, *which makes* 69; *subtract it from* 140, 71 *remains, and this is what the fifth possesses.*

The solution therefore proceeds as follows. In the system

$$\begin{cases} x_1 + x_2 + y = 2(x_3 + x_4 + x_5) \\ x_2 + x_3 + y = 3(x_4 + x_5 + x_1) \\ x_3 + x_4 + y = 4(x_5 + x_1 + x_2) \\ x_4 + x_5 + y = 5(x_1 + x_2 + x_3) \\ x_5 + x_1 + y = 6(x_2 + x_3 + x_4), \end{cases}$$

we add to each equation, as usual, the sum of the three unknowns appearing in the right-hand side to obtain S. This yields the system

$$\begin{cases} x_3 + x_4 + x_5 = \frac{1}{3}(S + y) \\ x_4 + x_5 + x_1 = \frac{1}{4}(S + y) \\ x_5 + x_1 + x_2 = \frac{1}{5}(S + y) \\ x_1 + x_2 + x_3 = \frac{1}{6}(S + y) \\ x_2 + x_3 + x_4 = \frac{1}{7}(S + y)\,. \end{cases}$$

Fibonacci sets $S + y = 420$, which is the lowest common multiple of the denominators. It then remains to solve

$$x_3 + x_4 + x_5 = 140 \tag{1}$$

$$x_4 + x_5 + x_1 = 105 \tag{2}$$

$$x_5 + x_1 + x_2 = 84 \tag{3}$$

$$x_1 + x_2 + x_3 = 70 \tag{4}$$

$$x_2 + x_3 + x_4 = 60. \tag{5}$$

As the sum of all five equations is 459, we have $S = 153$, and thus $y = 267$. Adding (2) and (4) yields $x_1 + S = 175$, so $x_1 = 22$. Next, by adding x_1

to (1), we obtain $S - x_2 = 162$. But $S = 153 < 162$, so "this problem is unsolvable, unless we put that the second man has a debt of 9, which is (the amount) from 153 to 162. So add together 22 deniers and the debt of the second, that is, subtract 9 from 22 (...)". The debt is thus a positive quantity which must be subtracted in the calculations involving x_2.

(3) Buying a Horse

A group of n people meet at a horse market. Each group of $t+1$ people, with t fixed ($0 \leq t \leq n-2$), taken cyclically starting with the jth person ($1 \leq j \leq n$), would like to buy a horse, but between them they do not have enough money; however, with a given fraction of the total sum possessed by the others, each of the groups attains the price of the desired horse. If x_l denotes the unknown assets of the lth person, y_j the price of the horse desired by the group beginning with the jth person, and m_j the given fraction, we have

$$\sum_{l=j}^{j+t} x_l + m_j \sum_{i \neq j,\ldots,j+t} x_i = y_j \qquad \text{for } j = 1, \ldots, n.$$

In the simplified case where all groups would like to buy the same horse, this becomes

$$\sum_{l=j}^{j+t} x_l + m_j \sum_{i \neq j,\ldots,j+t} x_i = y \qquad \text{for } j = 1, \ldots, n.$$

This problem differs from the previous one only by an exchange of terms. The solution formula can thus be obtained simply by changing the signs of the m's and y's in the formulas for the discovery of a purse:

$$\sum_{l=j}^{j+t} x_l = \frac{1}{n-(t+1)} \sum_{i=1}^{n} \frac{S - y_i}{1 - m_i} - \frac{S - y_j}{1 - m_j}$$

and

$$\sum_{l=j}^{j+t} x_l = (S - y) \left[\frac{1}{n-(t+1)} \sum_{i=1}^{n} \frac{1}{1 - m_i} - \frac{1}{1 - m_j} \right].$$

Such problems, but in pure numbers, can be found in Book I of Diophantus (I.24–25; see page 115), as well as, this time involving concrete quantities, in Chinese, Arab, and Byzantine mathematics[98]. In Abū Kāmil's *Algebra*, the purchase is already given to be a horse (Arabic *dābba*).

(4) The Disloyal Partners

Another type of linear system that Fibonacci studied also appears in Abū Kāmil's *Algebra*; as the latter refers to the solution method of other mathematicians, this type of problem must have been fairly well known, possibly even dating back to Greek antiquity. Its subject, and the origin of our name for this type, is the following. A group of men together have a certain amount of money, locked in a chest, of which each possesses a known

[98] Tropfke, *op. cit.* (note 69), pp. 608–9.

fraction. Each man holds a key to the chest. Bearing in mind Vergil's verse (*Aeneid* II.49) in O. Naises' version, *Quidquid id est, timeo socios et bona ferentes*, the men would do well to be suspicious of one another—indeed, while the others are away, each in turn helps himself from the chest, with the last leaving it completely empty. As each man's guilt is clear in the other men's minds, and is made even clearer by denials far too vehement to be sincere, our men finally agree that each would return a given fraction of what he stole and that the resulting sum would be divided equally among them. At the end of this distribution, the men find themselves in the initial situation, each having recovered his share.

In equations, these conditions can be expressed as

$$m_j x_j + \frac{1}{n}\sum_{i=1}^{n}(1-m_i)\,x_i = p_j S \qquad \text{for } j = 1,\ldots,n,$$

where S denotes the total amount, $p_j S$ (with $\sum p_i = 1$) the share belonging to the jth participant, x_j (with $\sum x_i = S$) the amount stolen by the jth, and $m_j x_j$ the amount he keeps (and thus $(1-m_j)\,x_j$ is the amount he returns).

To follow the same line as above, we will solve this system generally; but in this case Fibonacci considers the equations in succession. We can write the proposed system as

$$m_j x_j + \frac{1}{n}S - \frac{1}{n}\sum_{i=1}^{n} m_i x_i = p_j S,$$

which yields

$$(*) \qquad x_j = \frac{p_j}{m_j}S - \frac{1}{nm_j}S + \frac{1}{nm_j}\sum_{i=1}^{n} m_i x_i.$$

By adding these n equations, we obtain

$$S = S\sum\frac{p_i}{m_i} - \frac{1}{n}S\sum\frac{1}{m_i} + \frac{1}{n}\left(\sum\frac{1}{m_i}\right)\left(\sum m_i x_i\right),$$

and thus

$$\sum m_i x_i = \frac{S}{\sum\frac{1}{m_i}}\left(\sum\frac{1}{m_i} + n - n\sum\frac{p_i}{m_i}\right).$$

By substituting this into (*), we obtain

$$x_j = \frac{S}{m_j\sum\frac{1}{m_i}}\left[p_j\sum\frac{1}{m_i} - \frac{1}{n}\sum\frac{1}{m_i} + \frac{1}{n}\sum\frac{1}{m_i} + 1 - \sum\frac{p_i}{m_i}\right],$$

which simplifies to

$$x_j = \frac{S}{m_j\sum\frac{1}{m_i}}\left[\sum_{i\neq j}\frac{p_j - p_i}{m_i} + 1\right].$$

We see that the amount stolen by the jth participant will be positive if we have

$$\sum_{i \neq j} \frac{p_j - p_i}{m_i} > -1.$$

As for the other kinds of system, this condition holds in almost all of Fibonacci's examples. We will now examine the single exception in the six examples of these disloyal partners.

Example 4. *Liber abaci*, last problem "on three men having a sum in common" (*de tribus hominibus pecuniam communem habentibus*) from Chapter XII.7 (*Scritti*, I, pp. 296–97)

The system to be solved is

$$\begin{cases} \frac{1}{2}x_1 + \frac{1}{3}\left(\frac{1}{2}x_1 + \frac{1}{3}x_2 + \frac{1}{6}x_3\right) = \frac{1}{2}S \\ \frac{2}{3}x_2 + \frac{1}{3}\left(\frac{1}{2}x_1 + \frac{1}{3}x_2 + \frac{1}{6}x_3\right) = \frac{2}{5}S \\ \frac{5}{6}x_3 + \frac{1}{3}\left(\frac{1}{2}x_1 + \frac{1}{3}x_2 + \frac{1}{6}x_3\right) = \frac{1}{10}S. \end{cases}$$

As S is not fixed, the problem is indeterminate; we can thus choose any value. Fibonacci takes $S = 470$. With this choice, the initial (and final) assets of each man are integral, namely 235, 188, and 47 respectively. He can now calculate the stolen amounts (also integral). First, he finds 326 and 174 for x_1 and x_2, respectively. But then the money stolen by the first two, 500, appears to be greater than the total amount S (*que cum insimul iunguntur faciunt plus de* 470). In our writing, $x_3 = -30$. Thus, says Fibonacci, either the problem cannot be solved as posed, or we may suppose that, in addition to his share of the total, the third partner put in the chest an additional sum of 30 that belonged to him—a sum that would then find its way into the pockets of the first two (*quare hec questio non potest solvi nisi solvatur cum aliqua propria pecunia tertii hominis*). In this interpretation, we see that the theft results in a *gain* for the first two but a *loss* for the third.

We now consider the return of part of the stolen money. According to the conditions of the problem, the first and the second return $\frac{1}{2}x_1 = 163$ and $\frac{1}{3}x_2 = 58$; now the third is supposed to return $\frac{1}{6}x_3 = -5$. Fibonacci interprets this numerical result as follows: unable to return any money, the third participant has the opportunity to recover some of his previous loss by taking 5 from what his untrustworthy companions just returned (*post hec primus posuit in commune* $\frac{1}{2}$ *ex hoc quod ceperat, secundus* $\frac{1}{3}$*; ex quibus positionibus tertius homo cepit* $\frac{1}{6}$ *de ipsa pecunia propria, quam socii habuerant*). With these 5 and his third of the money returned, 72, he recovers both his share of the capital and his own money. Here we see that what the first two *give back* is reduced by what the third *takes*.

As we have seen, Leonardo Fibonacci did not always need to come up with such imaginative situations to interpret his negative answers. In other types of linear systems yielding negative solutions, such as that of the purse seen above, the participant whose assets take a negative value is said to have a debt, and the (positive) amount of this debt is then subtracted from the assets of the others wherever it would otherwise be added.

Yet in every case—about ten examples in his work as a whole—Leonardo Fibonacci declares the problem to be impossible unless one adopts the interpretation he proposes, which amounts to transforming the original concept (assets, gain) into the opposite concept (debt, loss), and, consequently, transforming a negative quantity into a positive quantity to be subtracted. Leonardo Fibonacci thus clearly refused to accept negative numbers. Nonetheless, by proposing such an interpretation, and by showing that in some cases this makes it possible to keep the proposed problem, he broke with the conventional practice of rejecting *a priori* a problem leading to a negative solution. Note also that such a change in attitude could only concern problems like the above systems, where all unknowns have a concrete signification and only one or two of them take a negative value. For only then could the negative solutions become meaningful within the framework of a real situation.

4.4. Later Developments

The Acceptance of Negative Numbers

We mentioned above (page 112) that two problems in Book I of Diophantus's *Arithmetica* are of the same type as those which consider the buying of a horse. These are the following two systems:

$$\begin{cases} x_1 + \frac{1}{3}(x_2 + x_3) = y \\ x_2 + \frac{1}{4}(x_3 + x_1) = y \\ x_3 + \frac{1}{5}(x_1 + x_2) = y \end{cases}$$

$$\begin{cases} x_1 + \frac{1}{3}(x_2 + x_3 + x_4) = y \\ x_2 + \frac{1}{4}(x_3 + x_4 + x_1) = y \\ x_3 + \frac{1}{5}(x_4 + x_1 + x_2) = y \\ x_4 + \frac{1}{6}(x_1 + x_2 + x_3) = y. \end{cases}$$

One might wonder why Diophantus decided to start his otherwise quite regular sequence of fractions with $\frac{1}{3}$ rather than $\frac{1}{2}$. Our formula from page 112 for the case $t = 0$,

$$(*) \qquad x_j = (S - y)\left[\frac{1}{n-1}\sum_{i=1}^{n}\frac{1}{1-m_i} - \frac{1}{1-m_j}\right],$$

shows why. If we started with $m_1 = \frac{1}{2}$ with the denominators being consecutive positive integers, we would obtain positive solutions only for $n = 3$ and $n = 4$. This is because, when $n = 5$,

$$\frac{1}{4}\sum_{i=1}^{5}\frac{1}{1-m_i} = \frac{437}{240} < \frac{1}{1-m_1} = 2,$$

and thus x_1 would be negative. If we start with $m_1 = \frac{1}{3}$, on the other hand, we obtain positive solutions up to $n = 6$. Remember what we have said (pages 33–34, 36) about Diophantus telling the reader to solve problems of the various kinds presented by himself. Because of this, it would have been unfair to trick him with a negative solution.

Some authors during the Middle Ages were not as considerate as Diophantus: such systems of two, three, or four equations beginning with $m_1 = \frac{1}{2}$ are common. Thus it is not surprising that a text on arithmetic, written around 1435 in the Provençal (Occitan) language by an unknown author in the town of Pamiers (South of France), considers the cases of three or four people wishing to buy a horse. What is new, however, is that this *Compendion del art de algorisme*, as this treatise is called, extends its study to the case $n = 5$ (which, for some reason, concerns the purchase of a piece of cloth instead), with the system

$$\begin{cases} x_1 + \frac{1}{2}(x_2 + x_3 + x_4 + x_5) = y \\ x_2 + \frac{1}{3}(x_3 + x_4 + x_5 + x_1) = y \\ x_3 + \frac{1}{4}(x_4 + x_5 + x_1 + x_2) = y \\ x_4 + \frac{1}{5}(x_5 + x_1 + x_2 + x_3) = y \\ x_5 + \frac{1}{6}(x_1 + x_2 + x_3 + x_4) = y. \end{cases}$$

The fact that this problem is indeterminate allows the author to arbitrarily fix some quantity. So taking $S - y = 60$, he first determines that

$$\frac{S-y}{1-m_i} = 120, \quad 90, \quad 80, \quad 75, \quad 72, \qquad \text{for } i = 1, \ldots, 5, \text{ respectively.}$$

He then calculates the values of the unknowns using formula (*) above: they are, respectively, $-\left(10+\frac{3}{4}\right)$, $19+\frac{1}{4}$, $29+\frac{1}{4}$, $34+\frac{1}{4}$, $37+\frac{1}{4}$. The "10 and $\frac{3}{4}$ that the first has less than nothing" does not seem to trouble the author, who does not offer any interpretation of this result. However, this is the only problem where he checks his treatment by substituting the values obtained for the unknowns into the proposed equations to prove that the same price y is indeed obtained in each case (of course with the assets of the first being taken as negative). This is the first known instance of accepting a negative quantity (and the only example of such a result in this Provençal treatise); the presence of verification and absence of interpretation show that this quantity is accepted for purely mathematical reasons.[99] [Appendix E.2 contains this problem, preceded by the two examples with $n = 3$ and $n = 4$]

Problems with negative solutions also appear in a 1484 work on algebra written in French in Lyons, Nicolas Chuquet's *Triparty*.[100] In some of these

[99] See J. Sesiano, "Une arithmétique médiévale en langue provençale," *Centaurus*, 27 (1984), pp. 26–75. (The publication of the complete text is forthcoming.)

[100] See A. Marre, "Notice sur Nicolas Chuquet et son Triparty en la science des nombres," *Bulletino di bibliografia e di storia delle scienze mathematiche e fisiche*, 13 (1880), pp. 555–659 and 693–814. The name of the work is due to its division into three parts.

problems, which involve pure numbers, the question of interpretation no longer even arises. This is the case for the following system, where one of the solutions is zero as well (a situation which was generally disregarded earlier; see pages 76, 79, and 107):

$$\begin{cases} x_2 + x_3 + x_4 + x_5 = 120 \\ x_3 + x_4 + x_5 + x_1 = 180 \\ x_4 + x_5 + x_1 + x_2 = 240 \\ x_5 + x_1 + x_2 + x_3 = 300 \\ x_1 + x_2 + x_3 + x_4 = 360. \end{cases}$$

This system is presented and solved in the *Triparty* as follows:

Next, I want to find five numbers such that all of them without the first make 120, *without the second* 180, *without the third* 240, *without the fourth* 300, *and without the fifth* 360.

And to find them I add together all these five numbers, and this makes 1200; *which I divide by* 4, *and I obtain* 300; *from which I subtract the five aforesaid numbers, namely* 120, 180, 240, 300, *and* 360, *and there remain to me* 180, 120, 60, 0, *and minus* 60, *which are the five numbers that I desired.*

Compared with some of Fibonacci's systems, this is a simple case, and solving it is a simple application of Thymaridas' rule: since the n equations are of the form $S - x_i = a_i$, we obtain, by addition, $(n-1)S = \sum a_i$, whence

$$x_j = S - a_j = \frac{1}{n-1} \sum a_i - a_j.$$

This is how Chuquet calculates (and how Fibonacci did, see his Example 3 above). Chuquet's original text is found in Appendix E.3, where this problem comes after some instructions about computing with negative numbers followed by a problem which is like the one above but leads to just the same values as the Provençal problem with the first negative solution accepted.

Even without considering this similar problem, it is likely, for both chronological and geographical reasons, that the *Triparty* descends from texts that are related, either directly or indirectly, to the *Compendion* from Pamiers. However, this does not imply that the idea of accepting negative answers originated in some school or tradition from the South of France. The time had simply come to give them a place in mathematics. The contemporary, and independent, presence of problems with negative answers in writings by the Tuscan Luca Pacioli, who is known to have collected knowledge and problems from several mathematicians he had met or read work of, confirms this[101]. Consider the following two examples of "ten problems"

[101] See the article on negative solutions mentioned in note 92. These problems are primarily found in an unpublished (and autograph) manuscript by Pacioli, written in 1476–80 for his students in Perugia (the shelf mark in the Vatican Library is Latin manuscript 3129—but only the dedication is in Latin).

(see page 62);

$$\begin{cases} x_1 + x_2 = 10 \\ 2x_1 + 3x_2 = 10 \end{cases}$$

and

$$\begin{cases} x_1 + x_2 = 10 \\ \frac{1}{2}x_1 + \frac{1}{3}x_2 = 10. \end{cases}$$

In the first system, Pacioli takes (our) x_1 to be the unknown x ("one thing", 1 *cosa*, abbreviated 1^{co}). He obtains from the second equation $x = 20$, and thus $x_2 = -10$ or, as he puts it, $x_2 = 10 - 20$, "which will satisfy the requirement" (*e noi ponemo che l'una parte fosse* 1^{co}, *doncha fo* 20, *et l'altra sirà* 10 *meno* 20, *e farà el quesito*). Doing the same in the second system, he obtains $x = 40$ and thus $x_2 = 10 - 40$ (*arai che l'una parte fo* 40, *l'altra fo* 10 *meno* 40). Another problem, involving concrete quantities, concerns the purchase of eggs with the following condition for the price: "Someone buys 7 eggs for as much less than 12 deniers as the 12 eggs cost him less than 8 deniers".[102] Pacioli then solves this. In our writing, since $7x = 12 - y$ and $y = 8 - 12x$, then $12 - 7x = 8 - 12x$, so $4 + 5x = 0$, whence $x = -\frac{4}{5}$. Pacioli says (see footnote) that the egg is "worth less" (*mensvalse*) or a "debt" (*indebitisse*). Here, however, the terms "debt" and "worth less" no longer actually express an interpretation, but instead are simply associated with the idea of a negative value.

The Cubic Equation

While at first algebra in medieval Europe was not significantly different from that in Muslim countries several centuries before, a transformation took place in Italy towards the end of the Middle Ages. This occurred in relation with the general solution of the cubic equation, which was the crucial step in the process separating algebra from geometry. However, who would have guessed, given the naïvety of early attempts to solve the cubic equation, that such an important development would soon arise? For example, in a document from 1328, Paolo Gerardi proposes solution formulas that follow from simply replacing the cubic equation by a quadratic equation, and, furthermore, doing so in the most arbitrary way imaginable[103]. Thus he states that the solution of $x^3 = ax^2 + bx + c$ is

$$x = \frac{a}{2} + \sqrt{\left(\frac{a}{2}\right)^2 + b + c},$$

[102] *Uno compra* 7 *ova per tanto men de* 12 *denari quanto che li costò li* 12 *ovi meno de* 8 *denari. Dimando che valse l'uovo. Dirai che l'uovo non valse, ma mensvalse e indebitisse* in the *Summa* (note 82), fol. 195^r; *7 ova vaglian tanto men de* 12 *denari quanto che* 12 *ova vaglian men de* 8 *denari. Dimando che valse l'uovo. Sapi che questa responde in mensvalere, altramente seria impossibile* in the manuscript, fol. 189^r.

[103] See W. van Egmont, "The Earliest Vernacular Treatment of Algebra: The *Libro di Ragioni* of Paolo Gerardi (1328)," *Physis*, 20 (1978), pp. 155–89.

which results from decreasing all the degrees of the unknown by one in the cubic equation; that is, he considers the equation $x^2 = ax + (b + c)$ instead. What seems to be mere fantasy turns out to be a method, as Gerardi uses a similar procedure to change a trinomial cubic equation into a trinomial quadratic equation, replacing $x^3 = bx+c$ by $x^2 = bx+c$ (here bx unchanged) and $x^3 = bx^2 + c$ by $x^2 = bx + c$. This at least allowed him to do with one less formula: the same applies to both resulting equations.

Even at the end of the fifteenth century, such attempts continued to be made. The manuscript by Luca Pacioli mentioned above solves (fol. 99^v and 260^r) $x^3 = ax^2 + c$ as

$$x = \left(\frac{a}{c}\right)^2 + c + \sqrt{\left(\frac{a}{2}\right)^2 + c} + \frac{a}{2}$$

and $x^3 + bx = c$ as

$$x = \sqrt{\left(\frac{b}{2}\right)^2 + b + c} - \frac{b}{2}.$$

In other instances, however, Pacioli is more prudent. In examining the "ten problem"

$$\begin{cases} x_1 + x_2 = 10 \\ x_1^2 \cdot x_2 = 143, \end{cases}$$

which leads to the equation $x_1^3 + 143 = 10x_1^2$, he declares (fol. 245^v) that no method for solving this equation is known, yet he hopes to find one (*sapi che finora el modo a questa (domanda) non s'è trovato, ma spero trovarlo*). On folio 150^r of his *Summa*, he remarks that when the constant term in an equation is associated with x and x^3, or x^2 and x^3, or x^3 and x^4, there are still no more known solution rules than for the problem of squaring the circle; there exist, however, empirical rules for some particular cases[104].

Nevertheless, during the two centuries leading to the Renaissance, a few attempts were made that proved to be less vain than these systematic reductions to quadratic equations. One example, from the second half of the fourteenth century, is that of Dardi of Pisa[105]. Since

$$(x + \alpha)^3 = x^3 + 3\alpha x^2 + 3\alpha^2 x + \alpha^3,$$

we have

$$x^3 + 3\alpha x^2 + 3\alpha^2 x = (x + \alpha)^3 - \alpha^3.$$

---

[104] *Ma de numero, cosa e cubo fra loro siando composti over de numero, censo e cubo, over de numero, cubo e censo de censo, non s'è possuto finora troppo bene formare regole generali (...) se non ale volte a tastoni (...) dirai che l'arte ancora a tal caso non à dato modo, sì commo ancora non è dato modo al quadrare del cerchio.*

[105] See W. van Egmont, "The Algebra of Master Dardi of Pisa," *Historia mathematica*, 10 (1983), pp. 399–421.

Then, by comparing this with $x^3 + ax^2 + bx = c$, Dardi sets

$$3\alpha = a$$
$$3\alpha^2 = b.$$

As $\alpha = \frac{b}{a}$, he deduces from $(x+\alpha)^3 - \alpha^3 = c$ the solution

$$x = \sqrt[3]{\left(\frac{b}{a}\right)^3 + c} - \frac{b}{a}.$$

Of course, this formula is only valid when $a^2 = 3b$ in the proposed equation. It is nonetheless innovative in its use of the expansion of the cube of the binomial, which leads to the presence of cube roots in the solution formula. This attempt to compare the expansion with the equation, which turned out to be of crucial importance later, was still remembered in the fifteenth century, as its inclusion in a *Libro d'abaco*, by the painter Piero della Francesca, shows[106]. This work also reproduces Dardi's analogous but less felicitous attempt at comparing

$$x^4 + ax^3 + bx^2 + cx = d$$

with

$$x^4 + 4\alpha x^3 + 6\alpha^2 x^2 + 4\alpha^3 x = (x+\alpha)^4 - \alpha^4$$

by setting

$$4\alpha = a,$$
$$\left(6\alpha^2 = b\right),$$
$$4\alpha^3 = c.$$

This implies that

$$\alpha^2 = \frac{c}{a},$$

so

$$\left(x + \sqrt{\frac{c}{a}}\right)^4 = \left(\frac{c}{a}\right)^2 + d$$

and

$$x = \sqrt[4]{\left(\frac{c}{a}\right)^2 + d} - \sqrt{\frac{c}{a}}.$$

Another notable attempt also dates back to the second half of the fourteenth century[107]. The root of the equation

$$x'^3 + ax'^2 = c$$

is said to be

$$x' = x - \frac{a}{3},$$

[106]See Piero della Francesca, *Trattato d'abaco*, edited by G. Arrighi, Pisa 1970, pp. 146-47.

[107]See R. Franci, "Contributi alla risoluzione dell'equazione di 3° grado nel XIV secolo," in *Mathemata* (Stuttgart 1985), pp. 221–28.

where x satisfies the equation

$$x^3 = \frac{1}{3}a^2x + \left[c - 2\left(\frac{a}{3}\right)^3\right].$$

Although we do not know if its importance was fully understood at the time, this transformation is fundamental. Indeed, if we set $y = x - \frac{a}{3}$ in the general equation

$$y^3 + ay^2 + by + c = 0,$$

we obtain a new equation, of the form

$$x^3 + px + q = 0,$$

which has no quadratic term. It is starting from this form that the general solution formula was established, a little more than a century later.

Chapter 5

Algebra in the Renaissance

5.1. The Development of Algebraic Symbolism

The end of the fifteenth century and the sixteenth century saw the birth of the first algebraic symbolisms. First, $\tilde{p}$ and $\tilde{m}$ were introduced in Italy as abbreviations for the Latin *plus* and the Italian *più* and for *minus* and *meno*, respectively, and were thus employed like our + and −. As we have remarked before (page 71), one of the key problems with verbal algebra had been the uncertainty about the span of the radical sign. This problem was settled as follows: ℞ or r (*radix*, "root") preceded a single term, while ℞.v (*radix universalis*, "collective root") announced that the root applied to a series of several terms, the end of which was then marked by the end of the mathematical expression or by some other break, such as a blank space. In 1572, all ambiguity finally disappeared with Bombelli's system of enclosing the radicand following R.q. or R.c., his symbols for square root and cube root, in the signs $\lfloor$ and $\rfloor$. Furthermore, the Latin words *res*, *census*, and *cubus* (or their Italian counterparts *cosa*, *censo*, and *cubo*) were replaced by abbreviations equivalent to symbols, as had happened in Greece. Sometimes these abbreviations were placed next to coefficients as a superscript, like exponents, for example in Pacioli (see his 1^{co} page 118). In Chuquet's *Triparty*, dating from the same time, it is the index of the power of the unknown which is placed next to its coefficient; thus $3x + 6x^2 = 30$ is expressed as 3^1 *plus* 6^2 *egaulx a* 30. Both of these systems of representation were used throughout the sixteenth century. Thus Bombelli would write the above expression as: $3^{\overset{1}{\smile}}$ *p.* $6^{\overset{2}{\smile}}$ *eguale à* 30.

Examples. Figure 32 is an excerpt from the end of Chapter XV of Cardano's *Ars Magna*[108]. It represents the numerical expressions

$$9 + \sqrt[3]{4846 + \frac{1}{2} + \sqrt{23\,487\,833 + \frac{1}{4}}} + \sqrt[3]{4846 + \frac{1}{2} - \sqrt{23\,487\,833 + \frac{1}{4}}}$$

[108] *Hieronymi Cardani, praestantissimi mathematici, philosophi, ac medici, Artis magnae, sive de regulis algebraicis, liber unus*, Nuremberg 1545 (reprinted in 1570 and 1663).

and

$$-\sqrt[3]{256+\frac{1}{2}+\sqrt{65\,063+\frac{1}{4}}}\ -\ \sqrt[3]{256+\frac{1}{2}-\sqrt{65\,063+\frac{1}{4}}}.$$

In Figure 33, Bombelli carries out an (awkwardly handled!) transformation of the equation $4+\sqrt{24-20x}=2x$ to determine the solution x (*il Tanto*, equivalent to *la cosa*)[109].

9. p̃. R̷. v. cub. $4846\frac{1}{2}$, p̃. R̷. $234878333\frac{1}{4}$,
p̃. R̷. v. cub. $4846\frac{1}{2}$ m̃. R̷. $234878333\frac{1}{4}$
m̃. R̷. v. cub. $256\frac{1}{2}$ p̃. R̷. $65063\frac{1}{4}$
m̃. R̷. v. cub. $256\frac{1}{2}$ m̃. R̷. $65063\frac{1}{4}$

FIGURE 32

Agguaglisi 4.p.R.q.L 24.m.20 $\overset{1}{\smile}$ J à 2 $\overset{1}{\smile}$ in simili agguagliamenti bisogna sempre cercare, che la R.q.legata resti sola, però si leuarà il 4.ad ambedue le parti, e si hauerà R.q.L 24.m.20 $\overset{1}{\smile}$ J. eguale à 2 $\overset{1}{\smile}$ m.4. Quadrisi ciascuna delle parti, si hauerà 24.m. 20. $\overset{1}{\smile}$ eguale à 4 $\overset{2}{\smile}$ m.16. $\overset{1}{\smile}$ p.16. lieuinsi li meni da ciascuna delle parti, e ponghansi dall'altra parte si hauerà 4 $\overset{2}{\smile}$ p.20 $\overset{1}{\smile}$ p.16. eguale à 24.p.16 $\overset{1}{\smile}$. lieuinsi li 16 $\overset{1}{\smile}$ à ciascuna delle parti, e si hauerà 4 $\overset{2}{\smile}$ p.4 $\overset{1}{\smile}$ p.16. eguale à 24. lieuisi il 16. da ogni parte si haueranno 4 $\overset{2}{\smile}$ p. 4 $\overset{1}{\smile}$ eguale à 8. riduchisi à 1 $\overset{2}{\smile}$ si hauerà 1 $\overset{2}{\smile}$ p.1 $\overset{1}{\smile}$ eguale à 2 (seguitisi il Capitolo) che il Tanto ualerà 1.

FIGURE 33

Algebraic symbolism was not developed only in Italy, and in fact the symbols now in use originated elsewhere. The symbols $+$ and $-$ appeared in Germany in the mid-fifteenth century (at least the origin of the first is clear: it was an abbreviated way of writing the Latin *et*, "and," commonly used for "plus"). The sign $\times$ is first found in W. Oughtred's *Clavis mathematicæ* (1631); the use of a dot instead, in order to avoid a confusion with x, was suggested by Leibniz in 1698. For division, Leibniz adopted $:$, while, earlier, the Swiss J. Rahn employed $\div$ (1659). The use of $\sqrt{}$ for roots comes from Germany (it may be a deformation of the letter r, the first letter of *radix*, as Euler believed). As for the index of the root, it was first denoted by writing the abbreviated name of the corresponding power underneath the radical

[109]See page 251 in the first edition, *L'Algebra, opera di Rafael Bombelli da Bologna, divisa in tre libri*, Bologna 1572 (it omits the last two books, which were published only in 1925 by E. Bortolotti), or page 192 in the complete edition of the work by E. Bortolotti, *L'Algebra, opera di Rafael Bombelli da Bologna*, Milan 1966.

sign, and later by the index itself, but initially also under the radical sign (note that Chuquet already wrote $\mathcal{R}^2$, $\mathcal{R}^3$, $\mathcal{R}^4$). Our symbol for equality appeared (alongside other contenders) concurrently in Bologna and England in the mid-sixteenth century; as R. Recorde writes when introducing this symbol in his *Whetstone of Witte* (London 1557), the use of a pair of parallel lines can be justified *bicause noe 2 thynges can be moare equalle.* Finally, let us mention that the symbolic use of letters was fixed by Descartes (1637), with the first and the last letters of the alphabet being used for known and unknown quantities, respectively.[110]

5.2. The General Solution of the Cubic and Quartic Equations

The Solution of the Cubic Equation

We can transform the general form of a cubic equation,

$$Ay^3 + By^2 + Cy + D = 0 \qquad \text{with } A,\ B,\ C,\ D \text{ real, } A \neq 0,$$

into the so-called normal form

$$y^3 + \frac{B}{A}y^2 + \frac{C}{A}y + \frac{D}{A} = 0,$$

or

$$y^3 + ay^2 + by + c = 0.$$

As the coefficients are real, this equation will have either one or three real roots.

Set $y = x + k$, where k will be chosen later. By replacing y by $x + k$ in the above equation, we obtain

$$\begin{aligned} &x^3 + 3kx^2 + 3k^2x + k^3 + ax^2 + 2akx + ak^2 + bx + bk + c \\ &\quad = x^3 + (3k + a)x^2 + (3k^2 + 2ak + b)x + (k^3 + ak^2 + bk + c) \\ &\quad = 0. \end{aligned}$$

Then choosing $k = -\frac{a}{3}$ yields

$$x^3 + \left(b - \frac{a^2}{3}\right)x + \left(\frac{2}{27}a^3 - \frac{ab}{3} + c\right) = 0,$$

or

$$x^3 + px + q = 0.$$

This is called the *reduced* form. The substitution $y = x - \frac{a}{3}$ thus allows us to transform an equation in normal form into an equation with no quadratic term.

On the other hand, the identity

$$(u + v)^3 = u^3 + 3u^2v + 3uv^2 + v^3$$

---

[110] An in-depth study of the development of algebraic symbolism can be found in Tropfke (see note 69) or the two very complete volumes by F. Cajori, *A History of Mathematical Notations*, Chicago 1928–29 (reprinted 1974).

can be written as

$$(u+v)^3 - 3uv(u+v) - (u^3+v^3) = 0,$$

and comparing this to the cubic equation shows that $u+v$ will be a solution if u and v can be determined so that

$$\begin{cases} u^3 + v^3 = -q \\ u \cdot v = -\frac{p}{3}. \end{cases}$$

This brings us back to an identity that we know well (pages 11, 66, 99). As

$$\frac{u^3+v^3}{2} = -\frac{q}{2},$$

while

$$\frac{u^3-v^3}{2} = \pm\sqrt{\left(\frac{q}{2}\right)^2 + \left(\frac{p}{3}\right)^3}$$

by the identity

$$\left(\frac{u^3+v^3}{2}\right)^2 = \left(\frac{u^3-v^3}{2}\right)^2 + u^3v^3,$$

we deduce that

$$u^3 = -\frac{q}{2} \pm \sqrt{\left(\frac{q}{2}\right)^2 + \left(\frac{p}{3}\right)^3}$$
$$v^3 = -\frac{q}{2} \mp \sqrt{\left(\frac{q}{2}\right)^2 + \left(\frac{p}{3}\right)^3}.$$

Since the roles of u and v may be interchanged, we will simply take the upper signs, so

$$u = \sqrt[3]{-\frac{q}{2} + \sqrt{\left(\frac{q}{2}\right)^2 + \left(\frac{p}{3}\right)^3}}$$
$$v = \sqrt[3]{-\frac{q}{2} - \sqrt{\left(\frac{q}{2}\right)^2 + \left(\frac{p}{3}\right)^3}}.$$

Thus

$$x = \sqrt[3]{-\frac{q}{2} + \sqrt{\left(\frac{q}{2}\right)^2 + \left(\frac{p}{3}\right)^3}} + \sqrt[3]{-\frac{q}{2} - \sqrt{\left(\frac{q}{2}\right)^2 + \left(\frac{p}{3}\right)^3}},$$

which is known today as *Cardano's formula*.

We know that if $z^3 = c$, then the cube root z can take the value $z_1 = \sqrt[3]{c}$ as well as two other values:

$$z_2 = \sqrt[3]{c} \cdot \frac{-1+i\sqrt{3}}{2}$$
$$z_3 = \sqrt[3]{c} \cdot \frac{-1-i\sqrt{3}}{2}.$$

The solutions x will thus be of the form $x = u_i + v_j$ $(i, j = 1, 2, 3)$, with

$$\begin{aligned} u_1 &= \sqrt[3]{-\tfrac{q}{2} + \sqrt{\left(\tfrac{q}{2}\right)^2 + \left(\tfrac{p}{3}\right)^3}}, & v_1 &= \sqrt[3]{-\tfrac{q}{2} - \sqrt{\left(\tfrac{q}{2}\right)^2 + \left(\tfrac{p}{3}\right)^3}} \\ u_2 &= u_1 \cdot \tfrac{-1+i\sqrt{3}}{2}, & v_2 &= v_1 \cdot \tfrac{-1+i\sqrt{3}}{2} \\ u_3 &= u_1 \cdot \tfrac{-1-i\sqrt{3}}{2}, & v_3 &= v_1 \cdot \tfrac{-1-i\sqrt{3}}{2}. \end{aligned}$$

From the condition $u_i \cdot v_j = -\frac{p}{3}$, however, there are only three possible combinations:

$$\begin{aligned} x_1 &= u_1 + v_1 \\ x_2 &= u_2 + v_3 \\ x_3 &= u_3 + v_2. \end{aligned}$$

The three roots of the cubic equation are thus

$$\begin{aligned} x_1 &= u_1 + v_1 \qquad \text{(which is Cardano's formula, as seen above)} \\ x_2 &= -\frac{u_1 + v_1}{2} + \frac{u_1 - v_1}{2} \cdot i\sqrt{3} \\ x_3 &= -\frac{u_1 + v_1}{2} - \frac{u_1 - v_1}{2} \cdot i\sqrt{3}. \end{aligned}$$

The number of real solutions is either one or three, depending on the sign of the discriminant.

(1) If the discriminant is *positive*—that is, if $\left(\frac{q}{2}\right)^2 + \left(\frac{p}{3}\right)^3 > 0$—then there are one real and two complex conjugate solutions.

(2) If the discriminant *equals zero*, then the three roots are real and the equation has a double root. Indeed, in this case we have

$$u_1 = v_1 = \sqrt[3]{-\frac{q}{2}} = -\sqrt[3]{\frac{q}{2}},$$

so

$$x_1 = -2\sqrt[3]{\frac{q}{2}},$$

and

$$x_{2,3} = \sqrt[3]{\frac{q}{2}}.$$

(3) Finally, if the discriminant is *negative*, then the three roots are real and distinct. Indeed, we can set

$$\begin{aligned} -\frac{q}{2} \pm \sqrt{\left(\frac{q}{2}\right)^2 + \left(\frac{p}{3}\right)^3} &= -\frac{q}{2} \pm i\sqrt{-\left[\left(\frac{q}{2}\right)^2 + \left(\frac{p}{3}\right)^3\right]} \\ &= r(\cos\varphi \pm i \sin\varphi), \end{aligned}$$

with

$$r = \sqrt{-\left(\frac{p}{3}\right)^3}$$

$$\cos\varphi = -\frac{q}{2r}$$

$$\sin\varphi = \frac{1}{r}\sqrt{-\left[\left(\frac{q}{2}\right)^2 + \left(\frac{p}{3}\right)^3\right]}.$$

By taking the cube root, we then obtain

$$u = \sqrt[3]{r}\left(\cos\frac{\varphi + k\cdot 2\pi}{3} + i\sin\frac{\varphi + k\cdot 2\pi}{3}\right)$$

$$v = \sqrt[3]{r}\left(\cos\frac{\varphi + k\cdot 2\pi}{3} - i\sin\frac{\varphi + k\cdot 2\pi}{3}\right),$$

where $k = 0, 1$, or 2 is chosen to be the same in u and v in order to satisfy the condition $u_i \cdot v_j = -\frac{p}{3}$. The roots of the equation are then

$$x_1 = 2\sqrt[3]{r}\cos\frac{\varphi}{3}$$

$$x_2 = 2\sqrt[3]{r}\cos\frac{\varphi + 2\pi}{3}$$

$$x_3 = 2\sqrt[3]{r}\cos\frac{\varphi + 4\pi}{3},$$

which are all real.

The Solution of the Quartic Equation

The normal form for the quartic equation,

$$x^4 + ax^3 + bx^2 + cx + d = 0,$$

can be written as

$$x^4 + ax^3 = -bx^2 - cx - d.$$

By adding $\left(\frac{1}{2}ax\right)^2$ to both sides, the left side becomes a square:

$$\left(x^2 + \frac{1}{2}ax\right)^2 = \left(\frac{1}{4}a^2 - b\right)x^2 - cx - d.$$

The left side remains a square if we add the expression $2(x^2 + \frac{1}{2}ax)y + y^2$ to both sides:

$$\left(x^2 + \frac{1}{2}ax + y\right)^2 = \left(\frac{1}{4}a^2 - b + 2y\right)x^2 + (ay - c)\,x + \left(y^2 - d\right).$$

Now the right side, which is of the form $Ax^2 + Bx + C$, will be a square as well if we require that $B^2 = 4AC$, because it can then be written as

$\left(\sqrt{A}x+\sqrt{C}\right)^2$. We thus need to set

$$(ay-c)^2-4\left(\frac{1}{4}a^2-b+2y\right)\left(y^2-d\right)=0$$

and solve this cubic equation in y. The (real) solution y_0 will then allow us to attain the desired form

$$\left(x^2+\frac{1}{2}ax+y_0\right)^2=\left(x\sqrt{\frac{1}{4}a^2-b+2y_0}+\sqrt{y_0^2-d}\right)^2,$$

and thus to write

$$x^2+\frac{1}{2}ax+y_0=\pm\left(x\sqrt{\frac{1}{4}a^2-b+2y_0}+\sqrt{y_0^2-d}\right).$$

Taking the sign to be positive yields one quadratic equation with solutions x_1 and x_2, and taking it to be negative yields a second quadratic equation with solutions x_3 and x_4. There are thus four roots, of which either none, two, or four are real.

5.3. The Solution of the Cubic Equation in Italy

First Solution Formula

We have seen that the equation with general form (in our notation)

$$y^3+ay^2+by+c=0$$

can be transformed via the substitution $y=x-\frac{a}{3}$ into a reduced equation having no quadratic term. As implied at the end of the last chapter, this was known, and applied, in the fifteenth century. Thus the various forms of the equation that may have a positive solution can be transformed into three cases of equations with positive coefficients—which are exactly analogous to the traditional three cases for the quadratic equation:

(1) $x^3+px=q$
(2) $x^3=px+q$
(3) $x^3+q=px$,

with p and q positive. The first two cases have a major advantage: there is always a positive solution (see page 88); even though negative solutions had been accepted earlier, this was preferably avoided. Yet it is the first case which is the most convenient to study since the discriminant is always positive. It was first solved around 1510 by Scipione dal Ferro (1465–1526), who had been teaching at the University of Bologna since 1496. Today we know only the statement of his solution formula, which was recorded in the sixteenth century by his disciple Pompeo Bolognetti and reads as follows: "When the things and the cubes are equal to the number, you are to reduce the equation to one cube by dividing by the quantity of cubes. Then cube the third part of the things, then square half the number, and add it to the aforesaid cubed. The square root of the above sum plus half the number makes a binomial, and the cube root of this binomial minus the cube root of the corresponding residue is the thing."[111]

The mathematical translation of this rule follows from knowing that, according to the ancient terminology, $\sqrt{m} - n$ ($\sqrt{m} > n$) is the "residue" of the "binomial" $\sqrt{m} + n$ while "the things" (*le cose*) denotes the coefficient of x (*la cosa*). The rule thus states that if the proposed equation is of the form $ax^3 + bx = c$, its reduced form $x^3 + px = q$ has solution

$$x = \sqrt[3]{\sqrt{\left(\frac{q}{2}\right)^2 + \left(\frac{p}{3}\right)^3} + \frac{q}{2}} - \sqrt[3]{\sqrt{\left(\frac{q}{2}\right)^2 + \left(\frac{p}{3}\right)^3} - \frac{q}{2}}.$$

While we have no information as to the derivation of this formula, there is little doubt that dal Ferro inferred it by comparing $(u-v)^3 + 3uv(u-v) = u^3 - v^3$ with $x^3 + px = q$. This immediately suggests setting $u - v = x$ and then determining u and v ($u > v$) that satisfy the system

$$\begin{cases} u^3 - v^3 = q \\ uv = \frac{p}{3}. \end{cases}$$

This is done by using the identity

$$\left(\frac{u^3 + v^3}{2}\right)^2 = \left(\frac{u^3 - v^3}{2}\right)^2 + u^3v^3,$$

just as we did above and as in Mesopotamian problems. As said, no difficulty arises with the discriminant since $p > 0$ and $q^2 > 0$.

Remark.

- Since $(u+v)^3 = 3uv\,(u+v) + u^3 + v^3$, dal Ferro could have solved $x^3 = px + q$ in the same way. He did not, probably because, as we infer by returning this equation to its general form $x^3 + px + q = 0$, we have $p < 0$ so that the discriminant may become negative.

Shortly before his death, dal Ferro passed this formula on to his student Antonio Maria Fior. This was an invaluable gift: in sixteenth-century Italy, a mathematician made a name for himself (as well as a living through the awarding of grants and posts) mostly by taking part in contests where mathematicians had to solve a proposed set of problems. Thus Fior decided to compete with the weapon provided by dal Ferro. In early 1535, he challenged the already well-known mathematician Niccolò Fontana (known as Tartaglia,

[111] *Dil cavaliero Bolognetti; lui l'hebbe da messer Sipion dal Ferro, vecchio bolognese. Il capitolo d(i) cose e cubi equale a numero.*
Quando le cose e li cubi si aguagliano al numer(o), ridurai la equatione a 1 *cubo partendo per la quantità delli cubi; poi cuba la terza parte de le cose, poi quadra la* $\frac{1}{2}$ *dil numero, e questo suma con il detto cubato, et la radice quadrata di deta summa più la* $\frac{1}{2}$ *dil numero fa un binomio; et la radice cuba di tal binomio men la radice cuba dil suo residuo val la cosa. Essempio:* 3 *cubi più* 10 *cose equale* 60. See Bortolotti (note 109), p. XVIII (text modified by us in accordance with the manuscript).

ca. 1500–1557) by proposing some thirty problems that led, not surprisingly, to an equation of the form $x^3 + px = q$. On February 12, 1535, eight days before the deadline for submitting the answers, Tartaglia finally managed to discover the solution formula, presumably in the same way as dal Ferro had. He was thus able not only to solve all of Fior's problems, but also to extend the solution method to the two remaining cases. So Tartaglia brilliantly won the contest and Fior's reputation as a mathematician was ruined for good. Yet fame sows the seeds of envy, as Tartaglia was about to discover.

The Controversy Between Tartaglia and Cardano

Girolamo Cardano was a highly renowned physician and mathematician, but also, later, one of ill repute, partly on account of envious colleagues and partly due to the peculiarity of his character. He himself, in his autobiographical *De propria vita*, admits to as many base shortcomings as worthy qualities. Like his character, his life was a strange mixture, alternating success and misfortune, for which he was, on the whole, entirely responsible. For instance, in one of his works he boasted of being able to cure pulmonary diseases. This earned him in 1552 a summons from the Scottish archbishop John Hamilton, whose illness had defeated even the most able physicians. And though he had probably actually cured no more asthmatics than lepers, Cardano set off for Scotland. We do not know whether it was by the power of suggestion or the talent of Cardano, but the archbishop's state improved to the extent that he recovered completely, and Cardano returned from Scotland with his head held high and his purse full. This favourable situation did not last: he had soon gambled away the money earned. That was nothing new, as his house had always been a gambling den attracting all sorts of disreputable characters. It is hardly surprising that the upbringing of his sons suffered accordingly. In 1560, the elder of the two was executed for poisoning his wife with arsenic—after, it is true, learning that he had had no part in the begetting of his own sons. (In all this turpitude, it is touching to note Cardano's desperate attempts to save him, his favorite son from the block.) On the other hand, it was Cardano himself who determined the fate of his younger son: tired of being openly defamed and secretly robbed, Cardano had him imprisoned and then expelled from the region of Bologna, cutting off his son's ear by way of farewell.

As well as gambling, Cardano practiced astrology. However, despite his predicting in London (while returning from Scotland) that the young King of England, Edward VI, would enjoy a long life, the latter was buried the following year. This did not bother Cardano, who simply revised his calculations to make the king's death agree with the will of the stars. Setting his sights steadily higher, Cardano then endeavoured to cast Jesus's horoscope. In this case his astrological knowledge definitely proved its limits, for he failed to predict his own condemnation by the Inquisition in 1570. He must, however, not have remained too long in prison, as he is known to have moved to Rome just a few months later. In the end, his great scientific reputation must have eclipsed his blasphemies, for he obtained a lifetime annuity from the Pope. It was in Rome that he died on September 20, 1576, after having allegedly starved himself for three days to ensure that he would die exactly on the day he had predicted.

So much for some of Cardano's exploits as physician and astrologist. It was in the realm of mathematics, however, that his reputation was the most distinguished as well as the most tarnished. In 1537, Cardano had already published a mathematical work, which, though not without value, was restricted to the exposition of the—hardly novel—theme of arithmetic and algebra limited to the first two degrees. He was already planning to write on the same subject a more complete *Ars Magna* when Giovanni Colla, who had himself tried to persuade Tartaglia to reveal his method, informed him during a trip to Milan of Tartaglia's triumph. Thus having learned that Tartaglia was now in possession of a method for solving cubic equations, Cardano begged the latter to entrust him with the solution for the case $x^3 + px = q$, by means of which he had solved Fior's problems. However Tartaglia was still smarting from having fallen for a similar plea by Colla who, once in possession of Tartaglia's solution to a few cubic problems, had claimed it as his own discovery. Therefore, not surprisingly, Tartaglia refused Cardano's request. But Cardano did not give up. Some time later, Cardano wrote to tell Tartaglia that the Marquis del Vasto, a man of considerable means and known for his liberality, was intent on meeting him. So Tartaglia traveled to Milan in March 1539, but failed to meet the marquis: not only had the latter been called away on business, but his servants knew nothing about the appointment. And so Tartaglia had no choice but to accept Cardano's hospitality; but Cardano refused to allow Tartaglia to leave before he had taught him his solving method. So Tartaglia eventually gave his host the short poem he had composed for himself to help memorize the solution formulas. Here is an English version of this poem and, as its meaning is not always obvious, a mathematical translation. See Appendix E.4 for the original.[112]

[112] *Quesiti et inventioni diverse de Nicolo Tartaglia*, fol. 120^v (1546; we have used the 1554 edition). Reproduced, along with various extracts of the epistolary dispute, in E. Bortolotti, "I contributi del Tartaglia, del Cardano, del Ferrari, e della scuola mathematica bolognese alla teoria algebrica delle equazioni cubiche," *Studi e memorie per la storia dell'Università di Bologna*, IX (1926), pp. 57–108. (Bortolotti is a fierce defender of Cardano's cause.)

When the cube, together with the things,	I. $x^3 + px = q$
is equated to some separate number,	
find two other numbers having it as their difference.	$u - v = q$
Then you are to hold to the rule	
that their product will always be equal	
to precisely a third (of the) cube of the things.	$u \cdot v = \left(\frac{p}{3}\right)^3$
Next, the ordinary corresponding remainder	
of their cube roots rightly subtracted	$x = \sqrt[3]{u} - \sqrt[3]{v}$
will be your main unknown.	
In the second of these manners,	II. $x^3 = px + q$
should the cube remain by itself,	
you must observe these other stipulations.	
You are to swiftly make of the number two parts	$u + v = q$
in such a way that one by the other produce directly	
a third (of the) cube of the things altogether.	$u \cdot v = \left(\frac{p}{3}\right)^3$
Of these then, by a usual procedure,	
you will take the cube roots, added together,	$x = \sqrt[3]{u} + \sqrt[3]{v}$
and this sum will be what you have in mind.	
The third of our present calculations	III. $x^3 + q = px$
is solved with the second, if you pay well attention,	[$x = -x_0$ if
for they are naturally, as it were, conjugate.	$x_0^3 = px_0 + q$].
I discovered these, and not at a slow pace,	
in one thousand five hundred and thirty-four,	
on quite sound and solid foundations,	
in the city surrounded by the sea.	

Remark.

- Tartaglia ends this mnemonic by giving as the place of his discovery Venice and as the date, 1534. This does not contradict his assertion that he had developed this method just before the competition with Fior (february 1535, see page 131), as in Venice the new year began in March.

Now Cardano had sworn on the Gospels never to divulge the contents of the poem, and even to write it in code so that no one would be able to understand it after his death.[113] In spite of this guarantee, Tartaglia eventually came to bitterly regret his weakness. Later letters from Cardano regarding some obscure passages of the poem (such as the "third cube") were first answered reluctantly, then Tartaglia replied in such a way as to deliberately lead Cardano astray, then finally he ignored his requests altogether. Yet Cardano continued his research, helped by his student Ludovico

[113] *Io vi giuro, ad sacra Dei evangelia, & da real gentil'huomo, non solamente da non publicar giamai tale vostre inventioni, se me le insignate. Ma anchora vi prometto, et impegno la fede mia da real Christiano, da notarmele in zifera, accio ché dapoi la mia morte alcuno non le possa intendere* (...). See the *Quesiti* (note 112), fol. 120^r.

Ferrari[114]. While they were staying in 1542 in Bologna, Cardano had the opportunity to consult dal Ferro's manuscripts, which had been kept by his son-in-law. Realizing that Tartaglia was not the initial discoverer, Cardano considered himself released from his oath. In 1545, he therefore published his *Ars Magna*, where, in addition to commonly known mathematics, he explains how to solve cubic equations, and also how to solve quartic equations, as discovered by Ferrari. Although he does make clear what he had learned from others and what his own contributions were,[115] this was meager consolation for Tartaglia. So Tartaglia continually accused Cardano publicly of perjury, which permanently damaged Cardano's reputation.

5.4. The Solution of the Quartic Equation in Italy

There is no question that the *Ars Magna* considerably developed what Cardano had learned from Tartaglia. First, each case of the cubic equation with positive terms, including quadrinomial expressions, is examined; next, following the traditional model for quadratic equations, geometric illustrations (this time with the help of stereometric diagrams) are included. Moreover, the *Ars Magna* contains an explanation of how to solve a quartic equation, which had been discovered by Ferrari while studying a problem proposed by Giovanni Colla to Cardano[116]. In our notation, the proposed system is

$$\begin{cases} a_1 + a_2 + a_3 = 10 \\ \dfrac{a_1}{a_2} = \dfrac{a_2}{a_3} \\ a_1 \cdot a_2 = 6. \end{cases}$$

By setting $a_2 = x$, we have $a_1 = \frac{6}{x}$ and $a_3 = \frac{x^3}{6}$, and the first equation then becomes

$$\frac{6}{x} + x + \frac{x^3}{6} = 10,$$

or, by multiplying both sides by $6x$,

$$x^4 + 6x^2 + 36 = 60x.$$

[114] 1522–1565; he was poisoned by his sister, a widow whom he had sheltered and who needed his estate to get married again.

[115] For example, at the beginning of Chapter XI (*De cubo & rebus equalibus numero*) of the *Ars Magna*, we read that *Scipio Ferreus Bononiensis iam annis ab hinc triginta fermè capitulum hoc invenit, tradidit verò Anthonio Mariae Florido Veneto, qui cùm in certamen cum Nicolao Tartalea Brixellense aliquando venisset, occasionem dedit, ut Nicolaus invenerit & ipse, qui cum nobis rogantibus tradidisset, suppressâ demonstratione, freti hoc auxilio, demonstrationem quæsivimus, eamque in modos, quod difficillimum fuit, redactam sic subiiciemus.* We find this same information at the very beginning of the book as well, except that Tartaglia is called *amicus noster*.

[116] *Fac ex* 10 *tres partes in continua proportione, ex quarum ductu primæ in secundam, producantur* 6. *Hanc proponebat Ioannes Colla, & dicebat solvi non posse, ego verò dicebam, eam posse solvi, modum tamen ignorabam, donec Ferrarius eum invenit* (*Ars Magna*, Chapter XXXIX, Problem V).

What comes next follows the general solution method we have already seen (page 128). First, we add $6x^2$ to both sides to obtain

$$\left(x^2+6\right)^2 = 6x^2 + 60x.$$[117]

Next, adding $2y(x^2+6)+y^2$ yields

$$\left(x^2+6+y\right)^2 = (2y+6)\,x^2 + 60x + \left(12y+y^2\right).$$

The right side will be a square if

$$900 = 2y^3 + 30y^2 + 72y,$$

that is, if

$$y^3 + 15y^2 + 36y = 450.$$

By setting $y = z - 5$, we obtain

$$z^3 - 39z - 380 = 0,$$

so, by Cardano's formula,

$$z = \sqrt[3]{190 + \sqrt{190^2 - 13^3}} + \sqrt[3]{190 - \sqrt{190^2 - 13^3}}.$$

Hence

$$y = \sqrt[3]{190 + \sqrt{33\,903}} + \sqrt[3]{190 - \sqrt{33\,903}} - 5.$$

We need hardly follow through this solution; the key point here is to give some idea of the unwieldiness of such solutions.

5.5. Bombelli and Imaginary Numbers

We have seen that in antiquity positive and rational solutions were preferred, that in medieval times computing with certain irrational numbers became more common, and that towards the end of the Middle Ages negative numbers began to be accepted—but only after it had been found that using them made sense in certain concrete situations. Under these circumstances, mathematicians would hardly have concerned themselves with the possibility of imaginary numbers. Yet such a situation was bound to arise sooner or later, particularly with certain "ten problems," since these continued to be a subject of predilection. Thus, among numerous systems of the same type in his manuscript (fol. 229^v–230^r), Pacioli considers the problem

$$\begin{cases} u + v = 10 \\ \frac{v}{u} + u = 5. \end{cases}$$

This leads to the equation $u^2 + 10 = 6u$, which is "impossible", he writes, "as the number is greater than the result of multiplying half of the x's by itself"

[117]The absence of a cubic term is not restrictive. In fact, this term may always be removed in the general (reduced) equation by the substitution $x = y - \frac{a}{4}$, where a is the coefficient of x^3 in the proposed equation.

(*perchè l'è più el numero che non è la multiplicatione dela* $\frac{1}{2}$ *dele chose in se. Impossibile.*).

Cardano's verdict is slightly different. In Chapter XXXVII of the *Ars Magna* he considers the problem

$$\begin{cases} u+v=10 \\ u\cdot v=40 \end{cases}$$

to be "clearly impossible" (*manifestum est quod casus seu quaestio est impossibilis*); nonetheless, he explicitly remarks that it would be satisfied by $5+\sqrt{-15}$ and $5-\sqrt{-15}$, for "25 minus the square of minus 15 is 40" (25 $\tilde{m}$ $\tilde{m}$ 15 *quad. est* 40).

Bombelli also shies away from accepting a solution to the equation $x^2+20=8x$, which he examines in his *Algebra* (note 109) when dealing with quadratic equations, among other examples of the form $x^2+q=px$. As he says, "this equality cannot hold except in the following, specious way" (*questo agguagliamento non si può fare se non in questo modo sofistico*), namely, to say that x equals $4+\sqrt{-4}$ or $4-\sqrt{-4}$. Nonetheless, later in that same work he himself introduces complex numbers. He then had good reason to be more favorably disposed towards their use in dealing with cubic equations, for it was the means to an end: that is, arriving at a real solution.

In looking at cubic equations, we have seen that when $p<0$ in the general (reduced) equation $x^3+px+q=0$—which corresponds to our types II and III, of the form $x^3=|p|x\pm|q|$—it may happen that $\left(\frac{q}{2}\right)^2+\left(\frac{p}{3}\right)^3<0$, that is, that the quantity under the square root is negative (even though type II always has one positive solution). This case of the negative discriminant is precisely the situation where the three roots of the equation are real and distinct. Now Bombelli had indeed remarked that the equation $x^3=15x+4$, although leading via Cardano's formula (with $p=-15$, $q=-4$) to the solution

$$\sqrt[3]{2+\sqrt{-121}}+\sqrt[3]{2-\sqrt{-121}},$$

was nevertheless satisfied by a positive solution, namely $x=4$. Moreover, he had geometrically demonstrated that *any* equation of the form $x^3=px+q$ must have a positive solution regardless of the size of the (here positive) values of p and q.

The general form of his proof, which he works out for the numerical example $x^3=6x+4$ [Appendix E.5], is as follows. Consider $x^3=px+q$, and draw two perpendicular lines on which take the known quantities $\mathrm{FL}=p$, $\mathrm{LM}=1$, and $\mathrm{LA}=\frac{q}{p}$ (see Figure 34); the area of the rectangle AF is then q.[118] Next, consider a right triangle whose apex I (at the right angle) moves along the line through A, L with one of its legs passing through the fixed point M and the other intersecting the line through L, F at a moving point

[118] As in ancient and medieval mathematics, rectangles are designated by the letters at two opposite angles (see pages 57, 68, 100).

G. Finally, consider a straight line starting at I and passing through F, which then intersects the line through A, B at some moving point C. It is clear that as the distance from I to L increases, the distance from G to F increases and the distance from C to B decreases; inversely, as the distance from I to L decreases, the distance from G to F decreases and the distance from C to B increases. Thus there must exist an intermediate configuration in which C and G are aligned vertically, namely that shown in Figure 35.

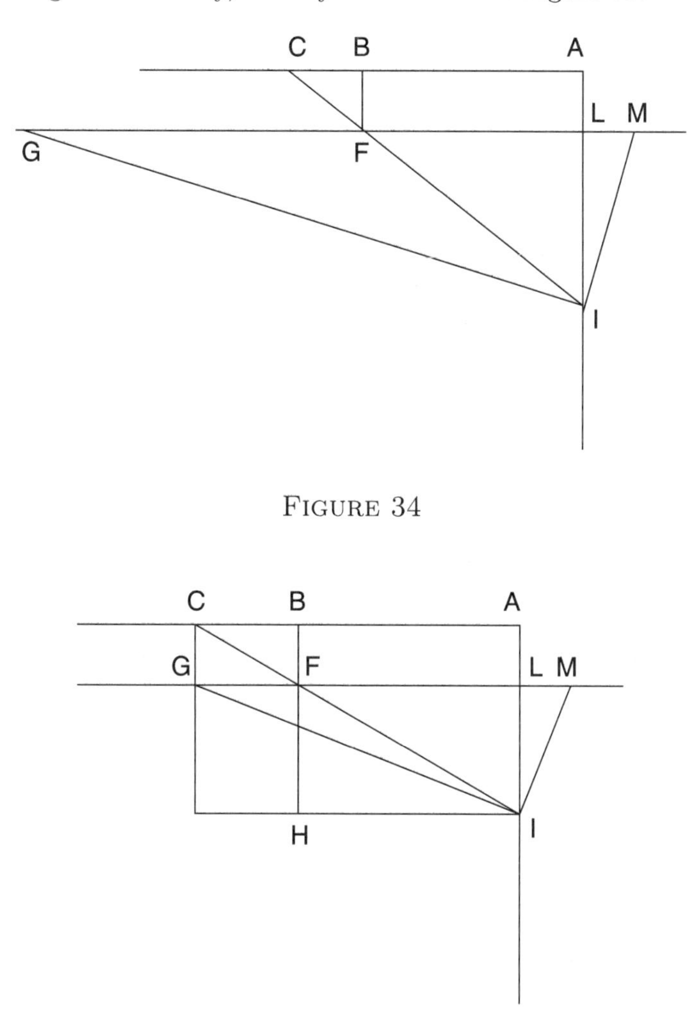

FIGURE 34

FIGURE 35

Now, consider this new figure and the rectangles it contains. First, the area of rectangle IG equals IL · LG; since (by the theorem on the height of a right triangle) $IL^2 = LG \cdot LM = LG$, the area of rectangle IG equals IL^3. Second, according to the theorem on complementary parallelograms (page 100), the two rectangles AF and GH have equal area, so that $GH = q$. Since the area of rectangle IG equals the sum of the areas of FI and GH, then, since $\mathrm{FL} = p$, the area of rectangle IG must also equal $p \cdot \mathrm{IL} + q$. It follows that $\mathrm{IL}^3 = p \cdot \mathrm{IL} + q$, and thus that IL is a positive solution of the equation $x^3 = px + q$.

The next step is to try to numerically express this positive solution. In his example of $x^3 = 15x + 4$ above, Bombelli sets, as we would write today,

$$\sqrt[3]{2 + \sqrt{-121}} = 2 + r\sqrt{-1}$$
$$\sqrt[3]{2 - \sqrt{-121}} = 2 - r\sqrt{-1},$$

which indeed yields the value $x = 4$ by addition. By cubing each side of the two equations, he then obtains, again in our writing,

$$2 + 11\sqrt{-1} = 8 + 12r\sqrt{-1} - 6r^2 - r^3\sqrt{-1}$$
$$2 - 11\sqrt{-1} = 8 - 12r\sqrt{-1} - 6r^2 + r^3\sqrt{-1},$$

and $r = 1$ follows by setting equal the real parts and the imaginary parts, respectively. He thus succeeds in obtaining the desired solution.

But, to do this, Bombelli had to apply new rules, specifically introduced by him for calculating products involving negative radicands, which departed from the usual product rule for square roots (with positive radicands), namely $\sqrt{k} \cdot \sqrt{l} = \sqrt{k \cdot l}$. Bombelli introduces these rules by saying that he has "found another kind of compound cube root, much different from the others, a kind which arises from the case of a cube equal to unknowns and a number when the cube of a third of the (coefficient of the) unknowns is larger than the square of half the number; (...) the computation with this kind of square root involves an operation different from the others and a different name."[119] The "different name" is "of minus" (*di meno*), meaning that the quantity below the radical sign is negative, as in $\sqrt{-k}$ (k positive), while "plus of minus" (*più di meno*) and "minus of minus" (*men(o) di meno*) indicate respectively that a plus and a minus sign precedes the root, as in $+\sqrt{-k}$ and $-\sqrt{-k}$. The new "operation" gave rise to the following four new rules:

- *più di meno via più di meno fa meno*, that is, $\left(+\sqrt{-k}\right)\left(+\sqrt{-k}\right) = -k$; this is, taking $k = 1$, our modern, $(+i)(+i) = -1$.
- *più di meno via men di meno fa più*, that is, $\left(+\sqrt{-k}\right)\left(-\sqrt{-k}\right) = +k$; equivalently, $(+i)(-i) = +1$.
- *meno di meno via più di meno fa più*, that is, $\left(-\sqrt{-k}\right)\left(+\sqrt{-k}\right) = +k$; equivalently, $(-i)(+i) = +1$.

[119]Edition by Bortolotti (note 109), pp. 133–34. These rules are introduced as follows: *Ho trovato un'altra sorte di R.c. legate molto differenti dall'altre, la qual nasce dal capitolo di cubo eguale a tanti e numero, quando il cubato del terzo delli tanti è maggiore del quadrato della metà del numero, (...) la qual sorte di R.q. ha nel suo algorismo* [=computation; see page 56] *diversa operatione dall'altre e diverso nome.* His *radice cubica legata* thus has the form $\sqrt[3]{u \pm \sqrt{-v}}$ and quantities of the form $\sqrt{-v}$ have a *nome diverso* and involve an *operatione diversa*.

- *meno di meno via men di meno fa meno*, that is, $\left(-\sqrt{-k}\right)\left(-\sqrt{-k}\right) = -k$; equivalently $(-i)(-i) = -1$.

However, Bombelli's troubles were not yet over, since generalization to the case where the positive solution is not already known (unlike in the above example) presented a difficulty. If one had

$$x = \sqrt[3]{u + \sqrt{-v}} + \sqrt[3]{u - \sqrt{-v}},$$

one could certainly set, as before,

$$\sqrt[3]{u + \sqrt{-v}} = s + \sqrt{-t}$$
$$\sqrt[3]{u - \sqrt{-v}} = s - \sqrt{-t},$$

with $s, t > 0$, so that the positive real solution would then be $x = 2s$. However, determining s as a function of the known quantities u and v requires solving an equation of the form $s^3 = \alpha s + \beta$—that is, one which is as problematic to solve as the original equation. Thus, since they were led to a like problem, mathematicians of the day called this the *casus irreducibilis*.

This first attempt to introduce and calculate with imaginary quantities was therefore limited; furthermore, it was purely utilitarian in that its sole purpose was to obtain real solutions; thirdly, it was in principle not to last for long, since it was soon to be rendered superfluous by trigonometric solution methods, which directly expressed all three real solutions without having to resort to imaginary quantities. Nevertheless, this latest and last extension of the domain of numbers was to play a role a few decades later, when the Fundamental Theorem of Algebra, which states that any equation of degree n has exactly n solutions, real or complex, was empirically found (though not to be proved for two centuries, by Gauss). Indeed, in his *Invention nouvelle en l'algebre*, published in 1629 in Amsterdam, A. Girard (1595–1632) already mentions that "all algebraic equations admit as many solutions as the denomination of the highest quantity shows" (*toutes les equations d'algebre reçoivent autant de solutions, que la denomination de la plus haute quantité le demonstre*). This is also noted by Descartes in the first pages of Book III of his *Géométrie* (published as an appendix to his *Discours de la méthode* in 1637), who at the same time introduces what was to become the name for these new numbers: "You are to know that in each equation, as many as the unknown quantity has dimensions, as many various roots can there be, that is, values of this quantity. (...) Moreover, both the true = positive) roots and the false (= negative) roots are not always real, but sometimes are only imaginary: that is, one can always imagine as many as I said in each equation, but sometimes there is no quantity that corresponds to those one imagines"[120]. It thus turns out that the new set of numbers suffices to express the solutions of any algebraic equation.

5.6. *Casus irreducibiles*

The successes in solving cubic and quartic equations led mathematicians of the following centuries to search for formulas yielding the solutions to higher degrees equations in the same way, that is, by means of an expression linking the coefficients in a finite number of applications of the four arithmetic operations and root extractions (so-called solutions "by radicals"). For, by analogy with the solutions of lower-degree algebraic equations, it was thought to be possible to find a transformation of the variable leading to a known solving method and root extractions.[121] Alternatively, an identity of the degree in question might allow one to establish a relation between the coefficients of the equation and the terms of the identity. With enough skill, it might even be possible to find a sequence of transformations eliminating the terms of intermediate degree and, ideally, yielding an equation of the form $x^n + k = 0$. The title of an article written by the inventor of the latter method, E. W. von Tschirnhaus (1651–1708), reflects this premature enthusiasm: "Method for removing all intermediate terms from a given equation"[122]. Indeed, little by little, mathematicians came to realize that for $n \geq 5$, the intermediate calculations required solving an equation of a degree higher than that of the original equation.

Thus it became less a question of how to solve equations of degree greater than four and more one of deciding if it was possible to solve them with a formula. The Italian P. Ruffini (1765–1822) was the first to prove—using an imperfect and complicated method—that it is impossible to solve a general equation of degree greater than four by radicals (*Teoria generale delle equazioni*, 1799). Later, the Norwegian N. Abel (1802–1829), who thought he had found a general solution for the fifth-degree equation (he did not know of Ruffini's work), was asked, before the printing of his discovery, to illustrate his theory by solving an example. This request proved beneficial,

[120] *Sçachés donc qu'en chasque Equation, autant que la quantité inconnue a de dimensions, autant peut il y avoir de diverses racines, c'est a dire de valeurs de cete quantité.* (...) *Au reste tant les vrayes racines que les fausses ne sont pas tousjours reelles, mais quelquefois seulement imaginaires: c'est a dire qu'on peut bien tousjours en imaginer autant que j'ay dit en chasque Equation, mais qu'il n'y a quelquefois aucune quantité, qui corresponde a celles qu'on imagine.*

[121] We have seen how the cubic equation is reduced to the second degree and root extractions, and how the quartic equation is reduced to solving a cubic and two quadratic equations. The solution of the quadratic equation is also formally reduced to a linear equation and a square root extraction: $x^2 + px + q = 0$ becomes $(x + \frac{p}{2})^2 + q = (\frac{p}{2})^2$, whence $y = (\frac{p}{2})^2 - q$ and a root extraction.

[122] "Methodus auferendi omnes terminos intermedios ex data aequatione," *Acta Eruditorum*, II (1683), pp. 204–7.

as vain attempts to do so led Abel to discover an error in his work and embark on the path which led to his proving the impossibility of solving the fifth-degree equation (*Mémoire sur les équations algébriques où on démontre l'impossibilité de la résolution de l'équation générale du cinquième degré*, published in Norway—in French—in 1824). It then remained to determine under what conditions a given equation could be solved by radicals. This question was settled by E. Galois (1811–1832), who in doing so developed concepts and methods that were fundamentally different from those used over the previous four millenia.

Appendix A

Mesopotamian Texts in Translation

(1)[123] Per bur, I obtained 4 kur of grain. Per second bur, I obtained 3 kur of grain. One grain exceeds the other by 8′20. I added my fields: 30′**0**. What are my fields?

Put 30′**0**, the bur. Put 20′**0**, the grain he obtained. Put 30′**0**, the second bur. Put 15′**0**, the grain he obtained. Put 8′20, that by which one grain exceeds the other. And put 30′**0**, the sum of the areas of the fields.

Next, divide in two 30′**0**, the sum of the areas of the fields: 15′**0**. Put 15′**0** and 15′**0**, twice. Take the inverse of 30′**0**, the bur: **0**.**0**′2. Multiply **0**.**0**′2 by 20′**0**, the grain he obtained: **0**.40, the false grain. Multiply by 15′**0**, which you have put twice: 10′**0**. Let your head hold. Take the inverse of 30′**0**, the second bur: **0**.**0**′2. Multiply **0**.**0**′2 by 15′**0**, the grain he obtained: **0**.30, the false grain. Multiply by 15′**0**, which you have put twice: 7′30. By what does 10′**0**, which your head holds, exceed 7′30? It exceeds by 2′30. Subtract 2′30, that by which it exceeds, from 8′20, that by which one grain exceeds the other; you leave 5′50. Let your head hold 5′50, which you left. Add the coefficient **0**.40 and the coefficient **0**.30: 1.10. I do not know its inverse. What must I put to 1.10 to have 5′50, that your head holds? Put 5′**0**. Multiply 5′**0** by 1.10; this gives you 5′50. From 15′**0**, which you have put twice, subtract, and to the other add, 5′**0**, which you have put; the first is 20′**0**, the second 10′**0**. The area of the first field is 20′**0**, the area of the second field 10′**0**.

If the area of the first field is 20′**0** and the area of the second field is 10′**0**, what is their grain? Take the inverse of 30′**0**, the bur: **0**.**0**′2. Multiply **0**.**0**′2 by 20′**0**, the grain he obtained: **0**.40. Multiply by 20′**0**, the area of the first field: 13′20, the grain of 20′**0**, the area of the first field. Take the inverse of 30′**0**, the second bur: **0**.**0**′2. Multiply **0**.**0**′2 by 15′**0**, the grain he obtained:

[123] 1 to 7 (Mesopotamian texts) from the French translation by Thureau-Dangin (n. 6). We separate sexagesimal digits by ′ and the integral and fractional parts by a dot. Initial and final zeros (non-existent in Mesopotamian clay tablets) are in bold.

0.30. Multiply **0**.30 by 10′**0**, the area of the second field: 5′**0**, the grain of 10′**0**, the area of the second field. By what does 13′20, the grain of the first field, exceed 5′**0**, the grain of the second field? It exceeds by 8′20.

(2) Per bur, I obtained 4 kur of grain. Per second bur, I obtained 3 kur of grain. Now two fields. One field exceeds the other by 10′**0**. One grain exceeds the other by 8′20. What are my fields?

Put 30′**0**, the bur. Put 20′**0**, the grain he obtained. Put 30′**0**, the second bur. Put 15′**0**, the grain he obtained. Put 10′**0**, that by which one field exceeds the other. And put 8′20, that by which one grain exceeds the other.

Take the inverse of 30′**0**, the bur: **0**.**0**′2. Multiply by 20′**0**, the grain he obtained: **0**.40, the false grain. Multiply **0**.40, the false grain, by 10′**0**, that by which one field exceeds the other: 6′40. Subtract from 8′20, that by which one grain exceeds the other; you leave 1′40. Let your head hold 1′40, that you left. Take the inverse of 30′**0**, the second bur: **0**.**0**′2. Multiply **0**.**0**′2 by 15′**0**, the grain he obtained: **0**.30, the false grain. By what does **0**.40, the false grain, exceed **0**.30, the false grain? It exceeds by **0**.10. Take the inverse of **0**.10, that by which it exceeds: 6. Multiply 6 by 1′40, which your head holds: 10′**0**, the area of the first field. To 10′**0**, the area of the field, add 10′**0**, that by which one field exceeds the other: 20′**0**. The area of the second field is 20′**0**.

If the area of the first field is 20′**0** and the area of the second field 10′**0**, what is their grain? Take the inverse of 30′**0**, the bur: **0**.**0**′2. Multiply **0**.**0**′2 by 20′**0**, the grain he obtained: **0**.40. Multiply by 20′**0**, the area of the field: 13′20, the grain of 20′**0**, the area of the field. Take the inverse of 30′**0**, the second bur: **0**.**0**′2. Multiply **0**.**0**′2 by 15′**0**, the grain he obtained: **0**.30. Multiply by 10′**0**, the area of the second field: 5′**0**, the grain of 10′**0**, the area of the field. By what does 20′**0**, the area of the field, exceed 10′**0**, the area of the second field? It exceeds by 10′**0**. By what does 13′20, the grain, exceed 5′**0**, the second grain? It exceeds by 8′20.

(3) I added the area and the side of my square: **0**.45.

You put 1, the unit. You divide in two 1: **0**.30. You multiply by **0**.30: **0**.15. You add **0**.15 to **0**.45: 1. It is the square of 1. You subtract **0**.30, which you multiplied, from 1: **0**.30, the side of the square.

(4) I subtracted from the area the side of my square: 14′30.

You put 1, the unit. You divide in two 1: **0**.30. You multiply by **0**.30: **0**.15. You add to 14′30: 14′30.15. It is the square of 29.30. You add **0**.30, which you multiplied, to 29.30: 30, the side of the square.

(5) A rectangle. I multiplied the length by the width, I thus constructed an area. Then I added to the area that by which the length exceeds the width: 3′3. Then I added the length to the width: 27. What are the length, the width, and the area?

27 3′3 the sums
15 the length
3′**0** the area
12 the width.

You proceed thus. Add 27, the sum of the length and the width, to 3′3: 3′30. Add 2 to 27: 29. Divide in two 29: 14.30; by 14.30: 3′30.15. From 3′30.15 you subtract 3′30: **0**.15, the remainder. **0**.15 is the square of **0**.30. Add **0**.30 to the first 14.30: 15, the length. You subtract **0**.30 from the second 14.30: 14, the width. You subtract 2, which you added to 27, from 14, the width: 12, the true width.

I multiplied 15, the length, by 12, the width. 15 by 12: 3.**0**, the area. By what does 15, the length, exceed 12, the width? It exceeds by 3. Add 3 to 3′**0**, the area: 3′3, (the sum of the excess of the length over the width and) the area.

(6) I added the area of my three squares: 23′20. The side of one exceeds the side of the other by 10.

You multiply by 1 the 10 that exceeds: 10. You multiply by 2: 20. You multiply 20 by 20: 6′40. You multiply 10 by 10: 1′40. You add to 6′40: 8′20. You subtract from 23′20: 15′**0**. You multiply by 3, the squares; you write 45′**0**. You add 10 and 20: 30. You multiply 30 by 30: 15′**0**. You add to 45′**0**: 1′**0**′**0**. It is the square of 1′**0**. You subtract 30, which you multiplied; you write 30. You multiply by 30 the inverse of 3, the squares, **0**.20: 10, the side of the square. You add 10 to 10: 20, the side of the second square. You add 10 to 20: 30, the side of the third square.

(7) I added the area of my two squares: 21′40. The side of one exceeds the side of the other by 10.

You divide in two 21′40; you write 10′50. You divide in two 10: 5. You multiply by 5: 25. You subtract from 10′50: 10′25. It is the square of 25. You write 25 twice. You add 5, which you multiplied, to the first 25: 30, the side of the square. You subtract 5 from the second 25: 20, the side of the second square.

Appendix B

Greek and Latin Texts

(1)[124] Ἐὰν ᾖ τρίγωνον ὀρθογώνιον οὗ ἡ μὲν κάθετος καὶ ἡ ὑποτείνουσα εἰς τὸ αὐτὸ ποδῶν η, ἡ δὲ βάσις ποδῶν δ, τούτου κατ᾽ ἰδίαν ζητήσομεν τήν τε κάθετον καὶ τὴν ὑποτείνουσαν.

Εὑρήσομεν δὲ οὕτως. Τὰ δ ἐφ᾽ ἑαυτά· γίνονται ιϛ· μέρισον εἰς τὸν η· γίνονται β· τὰ β ἄφελε ἀπὸ τῶν η· λοιπὰ ϛ· ὧν ἥμισυ γ. Ἔσται ἡ κάθετος γ. Ἔπειτα τὰ γ ἄφελε ἀπὸ τῶν η· λοιπὰ ε. Ἔσται ἄρα ἡ ὑποτείνουσα ποδῶν ε.

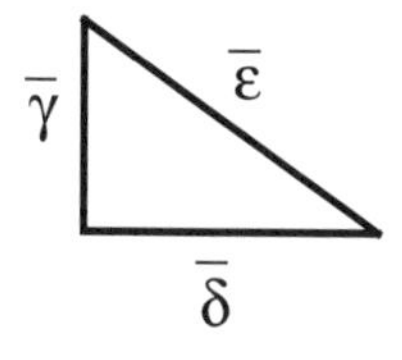

Ἐὰν ᾖ τρίγωνον ὀρθογώνιον οὗ ἡ μὲν κάθετος καὶ ἡ βάσις εἰς τὸ αὐτὸ ποδῶν ιζ, ἡ δὲ ὑποτείνουσα ποδῶν ιγ, εὑρεῖν τήν τε κάθετον κατ᾽ ἰδίαν καὶ τὴν βάσιν.

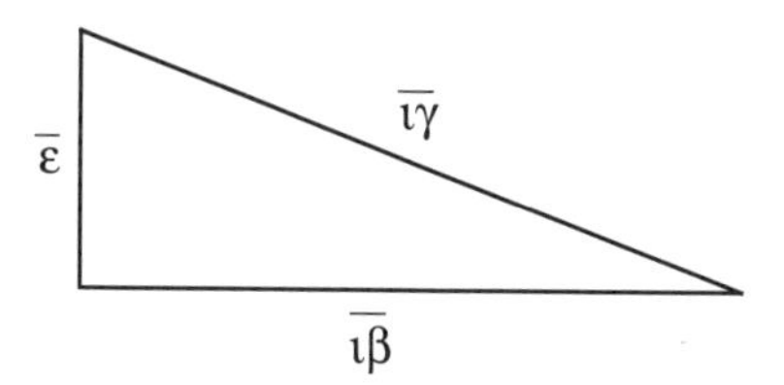

Εὑρήσομεν δὲ οὕτως. Τὰ ιγ ἐφ᾽ ἑαυτά, ρξθ. Καὶ τὰ ιζ ἐφ᾽ ἑαυτά· γίνονται σπθ. Τὰ ρξθ ποίησον δίς· γίνονται τλη. Καὶ τὰ σπθ ἄφελε ἀπὸ τῶν τλη· λοιπὰ μθ· ὧν πλευρὰ ζ. Ταῦτα ἄφελε ἀπὸ τῶν ιζ· λοιπὰ ι· τούτων τὸ ἥμισυ ε. Ἔσται ἡ κάθετος ε. Ταῦτα ἄφελε ἀπὸ τῶν ιζ· λοιπὰ ιβ. Ἔσται ἄρα ἡ βάσις ποδῶν ιβ.[125]

---

[124] Text from the edition cited in n. 20.

[125] The figure has been added by us on the assumption that such a figure appeared in the now missing part of the papyrus.

(2)[126] In trigono orthogonio, cuius hypotenusæ podismus est pedum XXV, embadum pedum CL, dicere cathetum et basim separatim.[127]

Sic quæritur. Semper multiplico hypotenusam in se, fit DCXXV. Ad hanc summam adiicio IIII embada, quæ faciunt pedes DC; utrumque in unum, fiunt pedes MCCXXV. Huius sumo latus, quod fit XXXV. Deinde ut interstitio duarum rectarum inveniatur: faciam hypotenusæ numerum in se, fit DCXXV; hinc tollo IIII embada, fiunt pedes XXV; huius sumo latus, fit V; erit interstitio. Hoc semper adiicio ad duas iunctas, id est ad XXXV, fiunt pedes XL. Huius sumo semper partem dimidiam, fiunt pedes XX. Erit basis trigoni. Si tollo de XX interstitionem, id est pedes V, reliqui sunt pedes XV. Erit cathetus eiusdem trigoni.

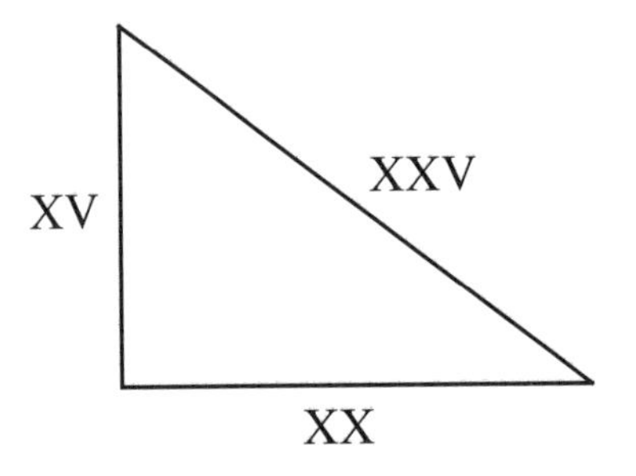

(3)[128] Τριγώνου ὀρθογωνίου τὸ ἐμβαδὸν μετὰ τῆς περιμέτρου ποδῶν σ̅π̅· ἀποδιαστεῖλαι τὰς πλευρὰς καὶ εὑρεῖν τὸ ἐμβαδόν.

Ποιῶ οὕτως. Ἀεὶ ζήτει τοὺς ἀπαρτίζοντας ἀριθμούς· ἀπαρτίζει δὲ τὸν σ̅π̅ ὁ δὶς τὸν ρ̅μ̅, ὁ δ̅ τὸν ο̅, ὁ ε̅ τὸν ν̅ϛ̅, ὁ ζ̅ τὸν μ̅, ὁ η̅ τὸν λ̅ε̅, ὁ ι̅ τὸν κ̅η̅, ὁ ι̅δ̅ τὸν κ̅. Ἐσκεψάμην, ὅτι ὁ η̅ καὶ λ̅ε̅ ποιήσουσι τὸ δοθὲν ἐπίταγμα.[129] Διὰ παντὸς λάμβανε δυάδα τῶν η̅· λοιπὸν μένουσιν ϛ̅ πόδες. Τὰ οὖν λ̅ε̅ καὶ τὰ ϛ̅ ὁμοῦ· γίνονται πόδες μ̅α̅. Ταῦτα ποίει ἐφ᾽ ἑαυτά· γίνονται πόδες ͵α̅χ̅π̅α̅. Τὰ λ̅ε̅ ἐπὶ τὰ ϛ̅· γίνονται πόδες σ̅ι̅. Ταῦτα ποίει ἀεὶ ἐπὶ τὰ η̅· γίνονται πόδες ͵α̅χ̅π̅. Ταῦτα ἆρον ἀπὸ τῶν ͵α̅χ̅π̅α̅· λοιπὸν μένει α̅· οὗ πλευρὰ τετραγωνικὴ γίνεται α̅. Ἄρτι θὲς τὰ μ̅α̅ καὶ ἆρον μονάδα α̅· λοιπὸν μ̅· ὧν ∠′[130] γίνεται κ̅· τοῦτό ἐστιν ἡ κάθετος, ποδῶν κ̅. Καὶ θὲς πάλιν τὰ μ̅α̅ καὶ πρόσθες α̅· γίνονται πόδες μ̅β̅· ὧν ∠′ γίνεται πόδες κ̅α̅· ἔστω ἡ βάσις ποδῶν κ̅α̅. Καὶ θὲς τὰ λ̅ε̅ καὶ ἆρον τὰ

---

[126] Text from Bubnov's edition (n. 23).

[127] Note the correspondence between Greek and Latin technical terms: *podismus* = ποδισμός, *embadum* = ἐμβαδόν, *cathetus* = κάθετος, *separatim* = κατ᾽ ἰδίαν, *in unum* = εἰς τὸ αὐτό, *latus* = πλευρά. *Trigonon*, *orthogonius*, *hypotenusa*, and *basis* are just transcriptions.

[128] Text from the edition cited in n. 27, pp. 422,15–424,5 (slightly emended by us). See also E. Bruins, *Codex Constantinopolitanus Palatii Veteris No. 1* (3 vol.), Leiden 1964, II, p. 39, and (reproduction of the Topkapı manuscript) I, pp. 56–57, (English translation) III, p. 90.

[129] We find here the interpolation: Τῶν σ̅π̅ τὸ η̅′· γίνονται πόδες λ̅ε̅. (η̅′ signifies $\frac{1}{8}$.)

[130] ∠′ stands for ἥμισυ.

$\overline{\text{ϛ}}$· λοιπὸν μένουσι πόδες $\overline{\text{κθ}}$· ἔστω ἡ ὑποτείνουσα ποδῶν $\overline{\text{κθ}}$. Ἄρτι θὲς τὴν κάθετον καὶ ποίησον ἐπὶ τὴν βάσιν· γίνονται πόδες $\overline{\text{υκ}}$· ὧν ∠′ γίνεται πόδες $\overline{\text{σι}}$. Τουτό ἔστι τὸ ἐμβαδόν. Καὶ αἱ τρεῖς πλευραὶ περιμετρούμεναι ἔχουσι πόδας $\overline{\text{ο}}$. Ὁμοῦ σύνθες μετὰ τοῦ ἐμβαδοῦ· γίνονται πόδες $\overline{\text{σπ}}$.

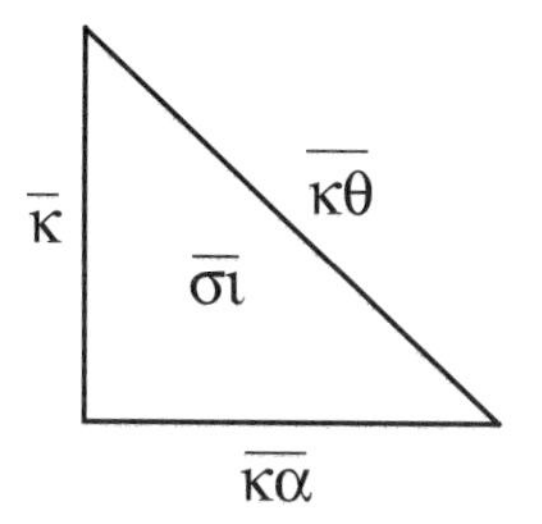

(4)[131] Τὸν ἐπιταχθέντα τετράγωνον διελεῖν εἰς δύο τετραγώνους.

Ἐπιτετάχθω δὴ τὸν $\overline{\text{ιϛ}}$ διελεῖν εἰς δύο τετραγώνους. Τετάχθω ἡ τοῦ πρώτου πλευρὰ ϛ $\overline{\text{α}}$, ἡ δὲ τοῦ ἑτέρου ϛ^{ῶν} ὅσων δήποτε ⋔ μονάδων ὅσων ἐστὶν ἡ τοῦ διαιρουμένου πλευρά· ἔστω δὴ ϛ $\overline{\text{β}}$ ⋔ M^{o} $\overline{\text{δ}}$. ἔσονται ἄρα οἱ $\square^{\text{οι}}$, ὃς μὲν Δ^{Υ} $\overline{\text{α}}$, ὃς δὲ Δ^{Υ} $\overline{\text{δ}}$ M^{o} $\overline{\text{ιϛ}}$ ⋔ ϛ $\overline{\text{ιϛ}}$. βούλομαι τοὺς δύο λοιπὸν συντεθέντας ἴσους εἶναι M^{o} $\overline{\text{ιϛ}}$. Δ^{Υ} ἄρα $\overline{\text{ε}}$ M^{o} $\overline{\text{ιϛ}}$ ⋔ ϛ $\overline{\text{ιϛ}}$ ἴσαι εἰσὶ M^{o} $\overline{\text{ιϛ}}$· καὶ γίνεται ὁ ϛ $\overline{\text{ιϛ}}^{\text{ε}}$.[132] ἔσται ἡ μὲν τοῦ πρώτου πλευρὰ $\overline{\text{ιϛ}}^{\text{ε}}$, αὐτὸς ἄρα ἔσται $\overline{\text{σνϛ}}^{\text{κε}}$, ἡ δὲ τοῦ δευτέρου πλευρὰ $\overline{\text{ιβ}}^{\text{ε}}$, αὐτὸς ἄρα ἔσται $\overline{\text{ρμδ}}^{\text{κε}}$. καὶ ἡ ἀπόδειξις φανερά.

(5) Τὸν δοθέντα ἀριθμὸν, ὃς σύγκειται ἐκ δύο τετραγώνων, μεταδιελεῖν εἰς δύο ἑτέρους τετραγώνους.

Ἔστω τὸν $\overline{\text{ιγ}}$, συγκείμενον ἔκ τε τοῦ $\overline{\text{δ}}$ καὶ $\overline{\text{θ}}$ τετραγώνων, μεταδιελεῖν εἰς ἑτέρους δύο τετραγώνους. Εἰλήφθωσαν τῶν προειρημένων τετραγώνων αἱ πλευραὶ M^{o} $\overline{\text{β}}$, M^{o} $\overline{\text{γ}}$, καὶ τετάχθωσαν αἱ τῶν ἐπιζητουμένων τετραγώνων πλευραί, ἣ μὲν ϛ $\overline{\text{α}}$ M^{o} $\overline{\text{β}}$, ἣ δὲ ϛ ὅσων δήποτε ⋔ μονάδων ὅσων ἐστὶν ἡ τοῦ λοιποῦ πλευρά. ἔστω ϛ $\overline{\text{β}}$ ⋔ M^{o} $\overline{\text{γ}}$· καὶ γίνονται οἱ τετράγωνοι, ὃς μὲν Δ^{Υ} $\overline{\text{α}}$ ϛ $\overline{\text{δ}}$ M^{o} $\overline{\text{δ}}$, ὃς δὲ Δ^{Υ} $\overline{\text{δ}}$ M^{o} $\overline{\text{θ}}$ ⋔ ϛ $\overline{\text{ιβ}}$. λοιπόν ἐστι τοὺς δύο συντεθέντας ποιεῖν M^{o} $\overline{\text{ιγ}}$. ἀλλ᾽ οἱ δύο συντεθέντες ποιοῦσιν Δ^{Υ} $\overline{\text{ε}}$ M^{o} $\overline{\text{ιγ}}$ ⋔ ϛ $\overline{\text{η}}$· ταῦτα ἴσα M^{o} $\overline{\text{ιγ}}$· καὶ γίνεται ὁ ϛ $\overline{\text{η}}^{\text{ε}}$.

Ἐπὶ τὰς ὑποστάσεις· ἔταξα τὴν τοῦ πρώτου πλευρὰν ϛ $\overline{\text{α}}$ M^{o} $\overline{\text{β}}$, ἔσται $\overline{\text{ιη}}^{\text{ε}}$, τὴν δὲ τοῦ δευτέρου πλευρὰν ϛ $\overline{\text{β}}$ ⋔ M^{o} $\overline{\text{γ}}$, ἔσται ἑνὸς^{ε}.[133] αὐτοὶ δὲ οἱ $\square^{\text{οι}}$ ἔσονται, ὃς μὲν $\overline{\text{τκδ}}^{\text{κε}}$, ὃς δὲ ἑνὸς^{κε}. καὶ οἱ δύο συντεθέντες ποιοῦσι $\overline{\text{τκε}}^{\text{κε}}$, ἃ συνάγει τὰς ἐπιταχθείσας $M^{o}$$\overline{\text{ιγ}}$.

(6) Τὴν μονάδα διελεῖν εἰς δύο μόρια καὶ προσθεῖναι ἑκατέρῳ τῶν τμημάτων τὸν δοθέντα ἀριθμὸν, καὶ ποιεῖν τετράγωνον.

---

[131] Text from Tannery's edition (n. 33), I, pp. 90–92 (B.4), 92–94 (B.5), 332–36 (B.6).

[132] That is, $\overline{\text{ιϛ}}$ πέμπτων.

[133] ἑνὸς^{ε} could also be written as $\overline{\text{ε}}$′.

Δεῖ δὴ τὸν διδόμενον μήτε περισσὸν εἶναι, μήτε τὸν διπλάσιον αὐτοῦ καὶ μονάδι μιᾷ μείζονα μετρεῖσθαι ὑπὸ τοῦ πρώτου ἀριθμοῦ οὗ ὁ μονάδι μιᾷ μείζων ἔχῃ μέρος τέταρτον.

Ἐπιτετάχθω δὴ ἑκατέρῳ τῶν τμημάτων προσθεῖναι M^{o} ϛ̅ καὶ ποιεῖν $\square^{ον}$. Ἐπεὶ οὖν θέλομεν τὴν μονάδα τεμεῖν καὶ ἑκατέρῳ τῶν τμημάτων προσθεῖναι M^{o} ϛ̅ καὶ ποιεῖν $\square^{ον}$, τὸ ἄρα σύνθεμα τῶν $\square^{ων}$ ἐστὶν M^{o} ι̅γ̅. δεήσει ἄρα τὸν ι̅γ̅ διελεῖν εἰς δύο $\square^{ους}$ ὅπως ἑκάτερος αὐτῶν μείζων ᾖ M^{o} ϛ̅. Ἐὰν οὖν τὸν ι̅γ̅ διέλω εἰς δύο $\square^{ους}$, ὧν ἡ ὑπεροχὴ ἐλάσσων ἐστὶν M^{o} α̅, λύω τὸ ζητούμενον· λαμβάνω τοῦ ι̅γ̅ τὸ ἥμισυ, γίνεται ϛ̅ ∠′, καὶ ζητῶ τί μόριον[134] προσθεῖναι M^{o} ϛ̅ ∠′ καὶ ποιεῖν $\square^{ον}$. καὶ πάντα τετράκις. ζητῶ ἄρα μόριον τετραγωνικὸν προσθεῖναι ταῖς κ̅ϛ̅ μονάσιν, καὶ ποιεῖν $\square^{ον}$. ἔστω τὸ προστιθέμενον μόριον $\Delta^{\Upsilon\times}$ α̅, καὶ γίνονται M^{o} κ̅ϛ̅ $\Delta^{\Upsilon\times}$ α̅ ἴσα $\square^{\varphi}$. Καὶ πάντα ἐπὶ Δ^{Υ}· γίνονται Δ^{Υ} κ̅ϛ̅ M^{o} α̅ ἴσα $\square^{\varphi}$· ἔστω τῷ ἀπὸ πλευρᾶς ς ε̅ M^{o} α̅, καὶ γίνεται ὁ ς M^{o} ι̅.[135] ἔσται ἄρα τὸ ταῖς κ̅ϛ̅ προστιθέμενον α̅^{ρ}, τὸ ἄρα ταῖς M^{o} ϛ̅ ∠′ καὶ γίνεται α̅^{υ} καὶ ποιεῖ $\square^{ον}$ τὸν ἀπὸ πλευρᾶς ν̅α̅^{κ}.[136]

Δεῖ οὖν τὸν ι̅γ̅ διαιρούμενον εἰς δύο $\square^{ους}$ κατασκευάζειν τὴν ἑκάστου πλευρὰν ὡς ἔγγιστα ν̅α̅^{κ}, καὶ ζητῶ τί ἡ τριὰς λείψασα, προσλαβοῦσα δυὰς ποιεῖ τὸν αὐτόν, τουτέστιν ν̅α̅^{κ}. Τάσσω οὖν δύο $\square^{ους}$, ἕνα μὲν ἀπὸ ς ι̅α̅ M^{o} β̅, τὸν δὲ ἕτερον ἀπὸ M^{o} γ̅ ⋀ ς θ̅ , καὶ γίνεται ὁ συγκείμενος ἐκ τῶν ἀπ᾽ αὐτῶν $\square^{ων}$ Δ^{Y} σ̅β̅ M^{o} ι̅γ̅ ⋀ ς ι̅ ἴσος M^{o} ι̅γ̅. καὶ γίνεται ὁ ς ε̅^{ρα}. ἔσται ἄρα ἑνὸς τῶν $\square^{ων}$ ἡ πλευρὰ σ̅ν̅ζ̅^{ρα}, ἡ δὲ τοῦ ἑτέρου σ̅ν̅η̅^{ρα}. καὶ ἐὰν ἀπὸ ἑκατέρου τῶν ἀπ᾽ αὐτῶν $\square^{ων}$ ἄρωμεν M^{o} ϛ̅, ἔσται τὸ μὲν ἓν τμῆμα τῆς μονάδος M^{o} ͵ε̅τ̅ν̅η̅$^{α.σα}$, τὸ δὲ ἕτερον ͵δ̅ω̅μ̅γ̅$^{α.σα}$, καὶ δῆλον ὡς ἑκάτερον μετὰ M^{o} ϛ̅ ποιεῖ $\square^{ον}$.

(7)[137] Propositio de episcopo qui iussit XII panes in clero dividi. Quidam episcopus iussit XII panes dividi in clero. Præcepit enim sic ut singuli presbyteri binos acciperent panes, diaconi dimidium, lector quartam partem, ita tamen fiat ut clericorum et panum unus sit numerus. Dicat qui valet quot presbyteri vel quot diaconi aut quot lectores esse debent.

Solutio. Quinquies bini fiunt X, id est, V presbyteri decem panes receperunt, et diaconus unus dimidium panem, et inter lectores VI habuerunt panem et dimidium. Iunge V et I et VI insimul, et fiunt XII. Rursusque iunge X et semis et unum et semis, fiunt XII. Hi sunt XII panes. Qui simul iuncti faciunt homines XII et panes XII; unus est ergo numerus clericorum et panum.

Propositio de emptore in C denariis. Dixit quidam emptor: Volo de centum denariis C porcos emere, sic tamen ut verres X denariis ematur, scrofa autem V denariis, duo vero porcelli denario uno. Dicat qui intelligit quot verres, quot scrofæ, quotve porcelli esse debeant ut in neutris nec superabundet numerus nec minuatur.

---

[134] Rather: τί μόριον τετραγωνικὸν.

[135] The text has here the gloss Δ^{Υ} γὰρ M^{o} ρ̅, τὸ $\Delta^{\Upsilon\times}$ M^{o} α̅^{ρ}.

[136] α̅^{ρ} and α̅^{υ} could also be written as ρ̅′ and υ̅′.

[137] Editions cited in n. 70, Problems 47 & 5.

Solutio de emptore. Fac VIIII scrofas et unum verrem in quinquaginta quinque denariis, et LXXX porcellos in XL, ecce porci XC; in quinque residuis denariis fac porcellos X, et habebis centenarium in utrisque numerum.

(8)[138] Cum volueris coquere mustum ignotum usque ad consumptionem duarum tertiarum, iam autem in coquendo consumptis de illo duabus mensuris et de remanenti effusis duabus mensuris, residuum vero in coquendo redactum est in duas mensuras et dimidiam, tunc quantum est mustum ignotum?

Sic facies. Numerum unde denominatur tertia, scilicet tres, multiplica in duas mensuras et dimidiam, et provenient septem et dimidia. Quibus adde duas consumptas et duas effusas, et fient undecim et dimidia. Deinde multiplica septem et dimidiam in duas consumptas, et fient quindecim. Deinde medietatem de undecim et dimidia, que est quinque et tres quarte, multiplica in se, et provenient triginta tres et dimidia octava. De quibus minue quindecim, et remanebunt decem et octo et dimidia octava. Cuius radici, que est quatuor et quarta, adde quinque et tres quartas, et fient decem; et tantum fuit mustum ignotum.

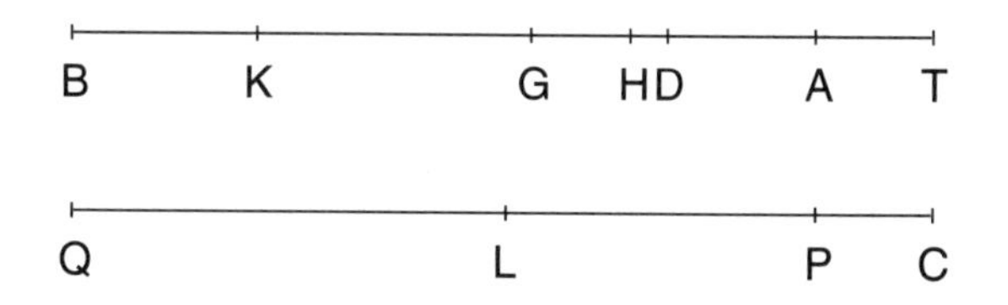

Cuius probatio est hec. Sit mustum ignotum linea AB, que[139] remanent de musto post consumptionem et effusionem pone lineam GB, duas vero mensuras in igne consumptas lineam AD, duas vero effusas lineam DG, duas vero et dimidiam in quas redigitur lineam KB. Iam autem diximus in eo quod precessit de musto cognito[140] quod talis est comparatio eius quod additur remanenti de musto post consumptionem duarum mensurarum ad hoc ut illud remanens cum addito fiat tantum ut eius tertia sit tertia totius musti ad idem remanens qualis est comparatio eius quod additur secundo remanenti post consumptionem duarum et post effusionem duarum aliarum ad hoc ut hoc secundum remanens cum addito fiat tantum ut eius tertia sit equalis ei quod queritur ad idem secundum remanens.[141] Constat igitur quia id quod additur linee BD quousque fiat linea AB sic se habet ad lineam BD

[138] Text of B.8 and B.9 from the edition in preparation (n. 85).

[139] *ae* becomes *e* in medieval Latin.

[140] In the preceding problems, the initial quantity of must was one of the givens.

[141] More simply, in our symbols:

$$\frac{3r_1 - (q-d)}{q-d} = \frac{3r_2 - (q-d-v)}{q-d-v}, \quad \text{or} \quad \frac{\text{AB} - \text{DB}}{\text{DB}} = \frac{3\text{BK} - \text{GB}}{\text{GB}}.$$

sicut id quod additur linee GB quousque tertia totius sit duo et dimidium quod est linea KB, ad BG. Lineam ergo additam linee GB ponemus lineam GH. Igitur linea KB est tertia linee BH. Igitur HB est septem et dimidium. Manifestum est igitur quod sic se habet linea AD ad lineam DB sicut se habet linea HG ad lineam GB. Componam autem proportionem; et talis erit comparatio linee AD ad lineam AB qualis est comparatio linee HG ad lineam HB.[142] Quod igitur fit ex ductu linee AD in lineam HB equum est ei quod fit ex ductu linee AB in lineam HG. Ex ductu autem linee BH in lineam AD proveniunt quindecim, quoniam linea AD est duo, linea vero HB est septem et dimidium. Unde cum multiplicatur linea HG in lineam AB provenient quindecim.[143] Deinde a puncto linee AB, scilicet a puncto A, protraham lineam equam linee HG; que est linea AT. Ex ductu igitur linee AT in lineam AB proveniunt quindecim.[144] Manifestum est autem quod linea HT est quatuor. Sed linea HB est septem et dimidium. Ergo linea TB est undecim et dimidium.[145] Faciam autem aliam lineam undecim et dimidii equalem linee TB; que est linea CQ. De qua incidam lineam equam linee TA; que est linea CP. Ex ductu igitur linee CP in lineam PQ proveniunt quindecim. Deinde dimidiabo lineam CQ in puncto L.[146] Quod igitur fit ex ductu linee CP in lineam PQ et linee PL in se equum est ei quod fit ex ductu linee CL in se, sicut dixit Euclides in libro secundo.[147] Sed ex ductu linee CL in se proveniunt triginta tres et dimidia octava, et ex ductu linee CP in lineam PQ proveniunt quindecim. Ergo ex ductu linee PL in se proveniunt decem et octo et dimidia octava. Ergo linea PL est quatuor et quarta. Linea vero QL est quinque et tres quarte. Igitur linea PQ est decem, et hoc est mustum incognitum.[148]

Vel aliter. Pone mustum ignotum rem. De qua re minutis duabus mensuris consumptis remanebit res minus duobus. De quo minue duas mensuras effusas, et remanebit res minus quatuor. Manifestum est igitur quod talis est comparatio rei minus duobus ad tertiam musti, que est tertia rei, qualis est comparatio rei minus quatuor ad duo et dimidium. Tantum igitur fit ex ductu rei minus duobus in duo et dimidium quantum ex ductu tertie rei in rem minus quatuor. Deinde fac sicut supra docuimus in algebra, et exit res decem.

---

[142] Putting $3\text{BK} = \text{HB}$, we have then $\text{AD} : \text{DB} = \text{HG} : \text{GB}$, whence (by composition of ratios)

$$\frac{\text{AD}}{\text{AD}+\text{DB}} = \frac{\text{HG}}{\text{HG}+\text{GB}}, \quad \text{that is,} \quad \frac{\text{AD}}{\text{AB}} = \frac{\text{HG}}{\text{HB}}.$$

[143] Since $\text{AB} \cdot \text{HG} = \text{AD} \cdot \text{HB}$ and $\text{AD} \cdot \text{HB} = 2(7 + \frac{1}{2}) = 15$, so $\text{AB} \cdot \text{HG} = 15$.

[144] Putting $\text{AT} = \text{HG}$, then $\text{AB} \cdot \text{AT} = 15$.

[145] Since $\text{HG} = \text{AT}$, so $\text{HT} = \text{AG} = 4$; but $\text{HB} = 7 + \frac{1}{2}$, so $\text{TB} = 11 + \frac{1}{2}$.

[146] Put $\text{CQ} = \text{TB}$, $\text{PC} = \text{AT}$. Then $\text{PQ}\cdot\text{PC} = 15$, $\text{CQ} = \text{PQ}+\text{PC} = 11+\frac{1}{2} = 2\cdot\text{CL}$.

[147] *Elements* II.5: $\text{PQ} \cdot \text{PC} + \text{PL}^2 = \text{CL}^2$.

[148] From $\text{PQ}\cdot\text{PC}$ and CL known, we find PL, whence we infer $\text{QL} + \text{PL} = \text{QP} = \text{AB}$.

(9) Si quis querat: Cum sint due diverse monete de quarum una dantur decem nummi pro morabitino, de altera vero triginta, cambit autem morabitinum aliquis pro nummis utriusque monete, sed subtractis hiis quos accipit de moneta decem nummorum pro morabitino de reliquis remanent viginti, tunc quot nummos accipit de utraque moneta[149]?

Sic facies. Agrega decem cum triginta, et fient quadraginta, quos pone prelatum[150]. Deinde minue viginti de triginta, et remanebunt decem; quos divide per prelatum, et exibit quarta, morabitini. Et hoc est quod accipit de moneta decem nummorum pro morabitino; sed tres quartas morabitini que remanent accipit de nummis alterius monete.

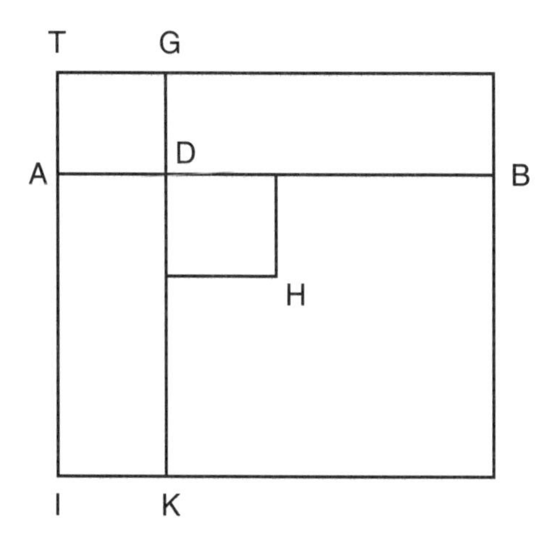

Monstrabitur autem hoc tali figura. Sit morabitinus linea AB, pars autem quam accipit de moneta decem nummorum pro morabitino sit linea AD, reliqua vero pars sit linea DB. Deinde de puncto D protrahatur linea de decem, que sit linea DG. Que multiplicetur in lineam DA, et proveniat superficies AG. Deinde de puncto D protrahatur linea de triginta, que sit linea DK. Que multiplicetur in lineam DB, et proveniat superficies KB. Deinde de superficie KB incidatur superficies equalis superficiei AG, que sit superficies DH [nam dixit quod subtractis nummis acceptis de nummis decem pro morabitino de nummis alterius monete remanent viginti][151]. Igitur superficies KBH, que remanet de maiore post incisionem superficiei DH, est viginti. Compleatur autem superficies IB.[152] Scimus autem quod linea AB est unum; nam ipsa est morabitinus. Linea vero AI est triginta. Igitur superficies IB est triginta. Monstratum est autem superficiem KBH esse viginti.[153] Igitur

[149] In view of the subsequent solution, the question should rather be: "quot partes morabitini accipit de utraque moneta?"

[150] Arabic *imām*, with the mathematical meaning of "divisor in the solution formula."

[151] This is clearly a former reader's remark incorporated into the text (and misplaced: it should have been copied after the next sentence).

[152] The upper part is completed as well.

[153] As stated just above.

restat ut due superficies DH et DI sint decem. Sed superficies DT equalis est superficiei DH. Igitur superficies IG est decem. Linea vero GK est quadraginta. Igitur per GK dividatur superficies IG, que est decem, et tunc linea TG erit quarta, que est equalis linee AD. Et hoc est quod accipit de moneta decem nummorum pro morabitino; et remanet DB tres quarte morabitini accipiende de nummis alterius monete. Et hoc est quod monstrare voluimus.

Vel aliter. Id quod accipit de moneta decem nummorum pro morabitino sit res. Id ergo quod accipit de moneta triginta nummorum pro morabitino est morabitinus minus re.[154] Multiplica igitur rem in decem, et provenient decem res. Deinde multiplica morabitinum minus re in triginta, et provenient triginta minus triginta rebus; de quibus minue decem res, et remanebunt triginta minus quadraginta rebus; que adequantur ad viginti. Comple[155] ergo triginta adiectis quadraginta rebus que desunt, et agrega totidem ad viginti; et fient triginta, que adequantur ad viginti et quadraginta rebus. Minue igitur viginti de triginta, et remanebunt decem; que adequantur quadraginta rebus. Res igitur est quarta. Et hoc est quod accipit de moneta decem nummorum pro morabitino, reliquum vero [quod remanet][156] accipit de nummis alterius monete. Vel, si volueris,[157] agrega decem res ad viginti, et fient viginti et decem res; que adequantur ad triginta minus triginta rebus. Fac igitur sicut supra docui in algebra, et erit res quarta. Et hoc est quod accipit de moneta decem nummorum pro morabitino, reliquum vero accipit de alia.

(10)[158] Quattuor homines habent denarios, ex quibus primus et secundus petunt reliquis 7 et proponunt habere ter tantum quam ipsi, secundus et tertius petunt reliquis 8 ut habeant quater tantum quam ipsi, tertius et quartus petunt reliquis 9 et habent quinquies tantum quam ipsi, quartus et primus petunt 11 et excedunt eos in sexcuplum; queritur quot unusquisque habeat.

Hec questio insolubilis est, et cognoscitur sic. Cum primus et secundus cum 7 ex denariis reliquorum habuerint ter tantum quam ipsi, tunc $\frac{3}{4}$ totius summe denariorum eorum habebunt ipsi, et tertio et quarto homini remanebit $\frac{1}{4}$ eiusdem summe.[159] Ergo inter tertium et quartum hominem habent $\frac{1}{4}$ totius summe, et amplius 7 quos dant primo et secundo homini.[160]

[154] That is, $1 - x$, whence the equation $30(1 - x) - 10x = 20$.

[155] Used in the same sense as "restaura", thus the Arabic *jabara*, applied to $30 - 40x = 20$.

[156] A reader must have considered *reliquum* to be unclear.

[157] Another way to deal with the equation $30(1 - x) - 10x = 20$.

[158] *Scritti* (n. 83) I, p. 201 (with minor corrections).

[159] From $x_1 + x_2 + 7 = 3(x_3 + x_4 - 7)$ we infer that $x_1 + x_2 + x_3 + x_4 = 4(x_3 + x_4 - 7)$, whence, with $S = x_1 + x_2 + x_3 + x_4$, $x_3 + x_4 - 7 = \frac{1}{4}S$ and $x_1 + x_2 + 7 = \frac{3}{4}S$. These transformations are known from previous problems.

[160] $x_3 + x_4 = \frac{1}{4}S + 7$.

Similiter ex petitionibus et ex propositionibus reliquorum invenies inter quartum et primum hominem habere $\frac{1}{5}$ totius summe et denarios 8, et inter primum et secundum $\frac{1}{6}$ dicte summe et denarios 9, et inter secundum et tertium $\frac{1}{7}$ eiusdem summe et insuper denarios 11. Et quoniam inter primum et secundum habent $\frac{1}{6}$ totius summe et denarios 9 et inter tertium et quartum $\frac{1}{4}$ eiusdem summe et denarios 7, ergo inter omnes quattuor habent $\frac{1}{6}\frac{1}{4}$ dicte summe et denarios 16.[161] Quare summa eorum est numerus de quo extracto $\frac{1}{6}\frac{1}{4}$, remanent 16; quem numerum (...) invenies esse $\frac{3}{7}$ 27.[162] Item quia inter quartum et primum habent $\frac{1}{5}$ totius summe eorum et denarios 8 et inter secundum et tertium habent $\frac{1}{7}$ et denarios 11, ergo erit summa eorumdem quattuor hominum quantum $\frac{1}{7}\frac{1}{5}$ eiusdem summe cum denariis 19.[163] Quare summa eorum est numerus de quo extracto $\frac{1}{7}\frac{1}{5}$ remanent 19; quem numerum (...) invenies esse $\frac{21}{23}$ 28. Quod est inconveniens cum per primam investigationem invenimus summam eorum esse aliter, scilicet $\frac{3}{7}$ 27. Unde hec questio insolubilis est.

Nam si eam solubilem proponere volumus, petant primus et secundus reliquis denarios 100, secundus et tertius denarios 106, tertius et quartus 145, quartus et primus 170, et invenies per utramque investigationem summam eorum esse 420; de qua inter primum et secundum habent $\frac{1}{6}$ et 145, scilicet 215, inter secundum et tertium habent $\frac{1}{7}$ de eodem 420 et 170, scilicet 230, et inter tertium et quartum habent $\frac{1}{4}$ de 420 et 100 plus, scilicet 205, et inter quartum et primum habent $\frac{1}{5}$ de 420 et denarios 106, hoc est 190. Quos divide inter eos ad libitum, hoc est, cum primus et secundus habent 215, habeat inde primus 100 et secundus 115; qui secundus, cum habeat cum tertio homine 230, extrahe inde 115, quos habet secundus, remanebunt tertio denarii 115; qui tertius, cum habeat cum quarto homine 205, extrahe inde 115 quos habet tertius, remanebunt quarto homini denarii 90.

(11)[164] De quinque hominibus et una bursa. Primus quidem et secundus habeant cum bursa duplum trium reliquorum hominum, secundus et tertius triplum, tertius et quartus quadruplum, quartus et quintus quincuplum, quintus et primus habeant similiter sexcuplum trium reliquorum hominum.

Ex hac quidem positione cognoscitur tertium et quartum et quintum hominem habere $\frac{1}{3}$ summe denariorum quinque hominum et burse, quartum quoque et quintum et primum $\frac{1}{4}$, quintum et primum et secundum $\frac{1}{5}$, primum et secundum et tertium $\frac{1}{6}$, secundum et tertium et quartum $\frac{1}{7}$.

Pone pro eorum summa et bursa 420, qui numerus dividitur integraliter per partes predictas, et accipe per ordinem $\frac{1}{7}$, $\frac{1}{6}$, $\frac{1}{5}$, $\frac{1}{4}$, $\frac{1}{3}$ ex ipsis. Et habebis

---

[161] $S = \frac{5}{12}S + 16$ ($\frac{1}{6}\frac{1}{4}$ means $\frac{1}{4} + \frac{1}{6} = \frac{5}{12}$).

[162] $\frac{3}{7}$ 27 means $27 + \frac{3}{7}$, which is S. This concludes the *prima investigatio* mentioned below.

[163] $S = \frac{1}{7}\frac{1}{5}S + 19$, that is, $S = \frac{12}{35}S + 19$, whence $S = 28 + \frac{21}{23}$ (*secunda investigatio*).

[164] *Scritti* I, pp. 227–28 (with minor corrections).

denarios tertii et quarti et quinti hominis 140, denarios quoque quarti et quinti et primi 105, quinti et primi et secundi 84, primi et secundi et tertii 70, similiter et denarii secundi et tertii et quarti erunt 60. Quibus quinque numeris insimul iunctis reddunt pro triplo denariorum quinque hominum 459, cum unusquisque ter computatus sit in prescriptis numeris.

Quare accipe $\frac{1}{3}$ de 459; cum cadat in integrum, exibunt 153 pro summa denariorum hominum. Qua extracta de 420 remanent 267 pro denariis burse. Post hec adde denarios primi et secundi et tertii cum denariis quarti et quinti et primi, scilicet 70 cum 105, erunt 175, et tot habent inter omnes, primo bis computato. Quare extrahe 153, scilicet summam eorum, de 175, remanent 22, et tot habet primus. Quos adde cum denariis tertii et quarti et quinti, erunt 162, et tot habent inter primum et tertium et quartum et quintum. Sed inter omnes quinque habent tantum 153. Quare hec questio est insolubilis nisi ponamus secundum hominem habere debitum 9, que sunt a 153 in 162. Adde itaque denarios 22 cum debito secundi, scilicet extrahe 9 de 22, remanent 13; quos extrahe de 70, remanent 57, et tot habet tertius; de quibus extrahe 9, scilicet debitum secundi, remanent 48; quos extrahe de denariis secundi et tertii et quarti, scilicet de 60, remanent 12, et tot habet quartus. Quos adde cum 57, erunt 69; quos extrahe de 140, remanent 71, et tot habet quintus.

Appendix C

Arabic Texts

(1)[165] فان قال عشرة قسمتها قسمين فقسمت هذا على هذا وهذا على هذا[166] فبلغ ذلك درهمين وسُدساً.

فقياس ذلك انّك اذا ضربت كلّ قسم فى نفسه ثمّ جمعتهما كان مثل احد القِسمين اذا ضربته فى الآخر ثمّ ضربت الذى اجتمع معك من الضرب فى الذى بلغ من القَسمين وهو اثنان وسُدس[167] فاضرب عشرة الّا شيئًا فى مثلها فتكون مائة ومالاً الّا عشرين شيئًا واضرب شيئًا فى شىء فيكون مالاً فاجمع ذلك فيصير مائة ومالين الّا عشرين شيئًا تعدل شيئًا مضروباً فى عشرة الّا شيئًا وذلك عشرة اشياء الّا مالاً مضروباً فى ما خرج من القَسمين وهو اثنان وسُدس فيكون ذلك احداً وعشرين شيئًا وثُلثى شىء الّا مالين وسُدساً[168] يعدل مائة ومالين الّا عشرين شيئًا فاجبر ذلك[169] وزد مالين وسُدساً على مائة ومالين الّا عشرين شيئًا وزد العشرين الشىء الناقصة من المائة والمالين على الواحد والعشرين الشىء وثُلثى الشىء فيكون معك مائة واربعة اموال وسُدس مال تعدل احداً واربعين شيئًا وثُلثى شىء فاردد ذلك الى مال واحد[170] وقد علمت انّ[171] المال الواحد من اربعة اموال وسُدس هو خُمسها

---

[165] Rosen's edition (n. 49), Arabic part pp. 32–33; text modified by us in places.

[166] The Latin version is clearer: *Divisi decem in duas partes et divisi hanc per illam et illam per istam.*

[167] If u and v are the two parts, then $u^2 + v^2 = uv(2 + \frac{1}{6})$.

[168] As in the next line, meaning وسُدس مال.

[169] Latin: *Restaura ergo illud* (meaning the two sides of $(21 + \frac{2}{3})x - (2 + \frac{1}{6})x^2 = 100 + 2x^2 - 20x$).

[170] The Arabic text has only مال, whereas we read in the Latin version: *Reduc ergo illud ad censum unum.* The subsequent reduction transforms $100 + (4 + \frac{1}{6})x^2 = (41 + \frac{2}{3})x$ into $24 + x^2 = 10x$.

[171] وقد علمت انّ = clearly (does not necessarily refer to anything that may have gone before)..

وخُمس خُمسها فخذ من جميع ما معك الخُمس وخُمس الخُمس فيكون معك اربعة وعشرون ومال تعدل عشرة اجذار[172] [لانّ العشرة[173] من احد واربعين شيئًا وثلثى شىء خُمسها وخُمس خُمسها][174] فنصّف الاجذار[175] وهى خمسة واضربها فى مثلها فتكون خمسة وعشرين فانقص منها الاربعة والعشرين التى مع المال يبق واحد فخذ جذره وهو واحد فانقصه من نصف الاجذار وهى خمسة فيبقى اربعة وهو احد القسمين.

(2)[176] فان قيل لك عشرة قسمتها قسمين فقسمت كلّ واحد على الآخر ثمّ ضربت كلّ واحد من القِسمين اللذين خرجا من قسمة كلّ واحد منهما على الآخر فى مثله والقيت الاقلّ من الاكثر فبقى درهمان.

قياسه ان نجعل ما خرج من قسمة الاقلّ على الاكثر شيئًا فتضربه فى مثله فيكون مالاً فتزيده على الدرهمين فيكون مالاً ودرهمين فبيّن ممّا وصفنا[177] ان ضَربَ ما يخرج من قسمة الكثير على القليل فى مثله مال ودرهمان فتضرب مالاً فى مال ودرهمين فيكون مال مال ومالين تعدل درهماً لانّا قد بيّنّا ان احدهما فى الآخر درهم[178] وضرب مربّع احدهما فى مربّع الآخر مثل الدرهم فى مثله وضرب الدرهم فى مثله درهم فنصّف المالين[179] فيكون واحداً فاضربه فى مثله فيكون واحداً فزده على الدرهم فيكون اثنين فجذر اثنين منقوص منه واحد مأخوذ جذر ما بقى[180] هو ما يخرج من قسمة القليل على الكثير.

ثمّ نرجع الى اوّل المسئلة[181] فنقول قسمنا عشرة دراهم الّا شيئًا على شىء فخرج جذر اثنين منقوص منه واحد مأخوذ جذر ما بقى فنضربه فى شىء فاذا اردت ذلك[182] فاضرب عشرة الّا شيئًا فى مثله فيكون مائة درهم ومالاً الّا عشرين

[172] *jidr* is synonymous with *shay'* (just as we speak of x or the *root* of an equation).

[173] *Sic*, without الشىء.

[174] The Latin omits this, which indeed looks like a reader's addition incorporated into the text.

[175] Halving the *coefficient* (with this begins the solution of the quadratic equation).

[176] Manuscript (n. 56), fol. 46^v–47^v (pp. 92–94 of the facsimile edition).

[177] In the statement.

[178] *dirham* = unit. That the product of two parts, one inverse of the other, is 1, has been used in previous problems. Here the two parts are x and $\frac{1}{x} = \sqrt{x^2+2}$, and the product of their squares is 1 as well.

[179] Solving the (bi)quadratic equation.

[180] Note the cumbersome expression to describe $\sqrt{\sqrt{2}-1}$ (overlapping radicals).

[181] The previous unknown (*shay'*, our x, which was $\frac{v}{u}$) has been determined. The same designation will now be used for a different unknown (namely $x = u$). Diophantus, who has just one designation for the unknown, does the same (n. 39).

[182] Long digression about this computation of $10 - u = u\sqrt{\sqrt{2}-1}$ ($u =$شىء).

شيئًا واضرب شيئًا فى مثله فيكون مالاً واضرب جذر اثنين منقوص منه واحد مأخوذ جذر ما بقى فى مثله فيكون جذر اثنين منقوص منه واحد ثمّ نقول قسمنا مائة درهم ومالاً الّا عشرين شيئًا على مال فخرج جذر اثنين منقوص منه واحد فاضربه فى مال فيكون جذر مالى مال منقوص منه مال يعدل مائة درهم و مالاً الّا عسرين شيئًا[183] فقابل به[184] وهو أن تجبر المائة درهم والمال بعشرين شيئًا وتزيدها على جذر مالى مال الّا مالاً وتجبر جذر مالى مال بمال وتزيده على مائة درهم ومال[185] وتلقى جذر مالى مال من مائة درهم ومالين فيكون مائة درهم ومالين الّا جذر مالى مال تعدل عشرين جذراً[186] فاردد كلّ شىء معك[187] الى مال وهو ان تضرب كلّ شىء معك فى واحد وجذر نصف واحد فتضرب مالين الّا جذر مالى مال فى درهم وجذر نصف درهم فيكون مالاً ثمّ تضرب عشرين شيئًا فى درهم وجذر نصف درهم فتكون عشرين شيئًا وجذر مائتى مال ثمّ تضرب مائة درهم فى درهم وجذر نصف درهم فتكون مائة درهم وجذر خمسة آلاف درهم فتجمع ذلك فيكون مال ومائة درهم وجذر خمسة آلاف درهم تعدل عشرين شيئًا وجذر مائتى مال[188] فنصّف عشرين شيئًا وجذر مائتى مال فيكون عشرة وجذر خمسين فاضربه فى مثله فيكون مائة وخمسين درهماً وجذر عشرين الفاً فانقص منه مائة درهم وجذر خمسة آلاف درهم فيبقى خمسون درهماً وجذر خمسة آلاف درهم فجذر ذلك منقوص من عشرة وجذرِ خمسين هو القسم الاكبر والقسم الاصغر باقى العشرة.

(3)[189] مسئلة. قيل ان انساناً أمر غلامه ان يتصدّق من ماله باثنى عشر درهماً على اثنى عشر نفراً وهم الرجال كلّ رجل بدرهمين والنساء كلّ امرأة بنصف درهم والصبيان كلّ صبىّ برُبع درهم.

الجواب. رجال خمسة بعشرة دراهم ونساء واحدة بنصف وصبيان ستّة بدرهم ونصف الجملة اثنا عشر نفراً باثنى عشر درهماً.

مسئلة. قيل ان انساناً دفع لوكيله مائة دينار وقال خذ لى مائة رأس بقر الثور بعشرة دنانير والبقرة بخمسة دنانير والعجل بنصف دينار.

---

[183] $\left(u^2(\sqrt{2}-1)=\right)\sqrt{2u^4}-u^2=100+u^2-20u$.

[184] *qābala* is also employed in the sense of reducing an equation to the standard form (thus performing the two operations of *al-jabr* and *al-muqābala*).

[185] Thus, after *al-jabr*: $\sqrt{2u^4}+20u=100+2u^2$.

[186] Thus, after *al-muqābala*: $100+(2u^2-\sqrt{2u^4})=20u$. What comes now is the reduction (making the coefficient of u^2 equal to 1).

[187] *shay'* is used here in the non-mathematical sense. See n. 213 (same expression).

[188] $u^2+(100+\sqrt{5000}\,)=(20u+\sqrt{200u^2}\,)$. What follows is the solution of this equation.

[189] Manuscript mentioned in n. 71.

الجواب انّه أخذ ثوراً واحداً بعشرة دنانير وبقرات تسع بخمسة واربعين وعجول
تسعين بخمسة واربعين الجملة مائة رأس بمائة دينار.

(4) [190] الصنف الاوّل هو مكعّب واضلاع تعدل عدداً.[191] نضع اب ضلع مربّع
مساوٍ لعدّة الجذور وهو مفروض ونعمل مجسّماً يكون قاعدته مثل مربّع اب ويكون
ارتفاعه مثل ب ج ويكون مساوياً للعدد المفروض[192] (...) ونجعل ب ج عموداً
على اب (...) ونخرج اب على استقامة الى ز ونعمل قطعاً مكافئًا رأسه نقطة ب
وسهمه ب ز وضلعه القائم اب وهو قطع ح ب د[193] فيكون قطع ح ب د معلوم
الوضع (...) ويكون ممّاساً لخطّ ب ج ونعمل على ب ج نصف دائرة فانّها باضطرار
تقطع القطع[194] فلتقطعه على د ونخرج من د التى[195] هى معلومة الوضع كما عرفته
عمودى دز ده على ب ز ب ج فيكونان معلومى الوضع والقدر فخطّ دز من خطوط
الترتيب فى القطع[196] فيكون مربّعه مساوياً لضرب ب ز فى اب[197] فيكون نسبة اب
الى دز الذى هو مثل ب ه كنسبة ب ه الى ه د الذى هو مثل زب[198] لكن نسبة
ب ه الى ه د كنسبة ه د الى ه ج فالخطوط الاربعة[199] متناسبة اب ب ه ه د
ه ج[200] فنسبة مربّع اب الاوّل الى مربّع ب ه الثانى كنسبة ب ه الثانى[201] الى ه ج
الرابع[202] فالمجسّم الذى قاعدته مربّع اب وارتفاعه ه ج مساوٍ لمكعّب ب ه[203]
(...) ونجعل المجسّم الذى قاعدته مربّع اب وارتفاعه ه ب مشتركاً[204] فيكون
مكعّب ب ه مع هذا المجسّم مثل المجسّم الذى قاعدته مربّع اب وارتفاعه ب ج[205]
الذى فرضناه مساوياً للعدد المفروض لكن المجسّم الذى قاعدته مربّع اب الذى هو

---

[190] This follows Woepcke's edition of the Arabic text (n. 74), pp. 20–21.

[191] First kind of the trinomial equations.

[192] Thus, for $x^3 + bx = c$, $\mathrm{AB} = \sqrt{b}$, $\mathrm{AB}^2 \cdot \mathrm{BG} = b \cdot \mathrm{BG} = c$, so $\mathrm{BG} = \frac{c}{b}$.

[193] رأس = vertex, سهم = axis, ضلع قائم = parameter (*latus rectum*).

[194] Otherwise, there would be no solution: D determines E (Arabic ه), and BE is x.

[195] In the feminine: D is a نقطة.

[196] خطّ الترتيب فى القطع is the ordinate of the (conic) section.

[197] $\mathrm{DZ}^2 = \mathrm{BZ} \cdot \mathrm{AB}$.

[198] $\mathrm{AB} : \mathrm{DZ} = \mathrm{DZ} : \mathrm{BZ}$, so (as $\mathrm{DZ} = \mathrm{BE}$, $\mathrm{BZ} = \mathrm{ED}$) $\mathrm{AB} : \mathrm{BE} = \mathrm{BE} : \mathrm{ED}$.

[199] There were three before.

[200] $\mathrm{BE} : \mathrm{ED} = \mathrm{ED} : \mathrm{EG}$, so $\mathrm{AB} : \mathrm{BE} = \mathrm{ED} : \mathrm{EG}$.

[201] Both times the second term in the previous proportion.

[202] $\frac{\mathrm{AB}^2}{\mathrm{BE}^2} = \frac{\mathrm{BE}}{\mathrm{ED}} \cdot \frac{\mathrm{ED}}{\mathrm{EG}} = \frac{\mathrm{BE}}{\mathrm{EG}}$.

[203] $\mathrm{AB}^2 \cdot \mathrm{EG} = \mathrm{BE}^3$.

[204] $\mathrm{AB}^2 \cdot \mathrm{EG} + \mathrm{AB}^2 \cdot \mathrm{BE} = \mathrm{BE}^3 + \mathrm{AB}^2 \cdot \mathrm{BE}$.

[205] $\mathrm{AB}^2 \cdot \mathrm{BG} = \mathrm{BE}^3 + \mathrm{AB}^2 \cdot \mathrm{BE}$.

مثل عدّة الجذور وارتفاعه ه ب الذى هو ضلع المكعّب مساوٍ لعدّة اضلاع مكعّب ه ب المفروضة فمكعّب ه ب مع عدّة اضلاعه المفروضة مساوٍ للعدد المفروض وذلك المراد.[206]

وليس لهذا الصنف اختلاف وقوع ولا يستحيل من مسائله شىء وقد خرج بخواصّ الدائرة مع خواصّ القطع المكافى.

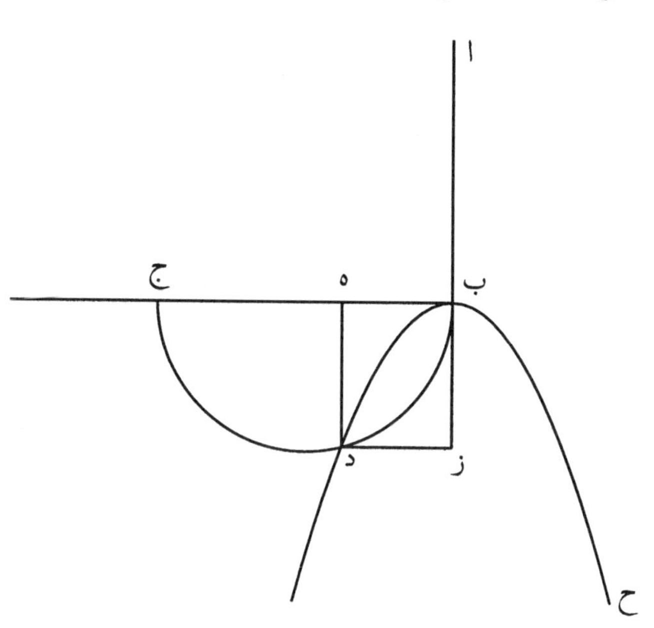

[206] Since $AB^2 = b$, thus $AB^2 \cdot BE = b \cdot BE$, while $AB^2 \cdot BG = c$, so $BE^3 + b \cdot BE = c$.

Appendix D

Hebrew Text[207]

ואם יאמרו לך עשרה בגדים בין שני אנשים בשכר שבעים ושנים
אדרהמיש וכל אחד מהם הוציא שלשים וששה אדרהמיש ואחד מהאנשים
לקח מהבגדים ולקח האחר הנשאר מהבגדים ושומת כל בגד מהאיש
האחד יותר משומת כל בגד מהאחר שלשה אדרהמיש.

ומעשיהו שנשים העשרה בגדים קו אב והבגדים שלקח אחד
מהם[208] קו אג והאחרים קו גב ונשים שומת כל בגד מקו גב קו גד וכל
שומת הבגדים מקו גב שטח דב ושטח דב הוא שלשים וששה ותשים שומת
כל בגד מקו אג קו גה וכל שומת הבגדים מקו גא יהיה שטח אה והוא
שלשים וששה וקו דה שלושה בעבור כי הוא יתרון שומת כל בגד מאחד
מהם על שומת כל בגד מהאחר ונשים קו גב דבר[209] והוא כמו קו דז וקו
דה שלשה ושטח הז יהיה מפני זה שלשה דברים וכל שטח בע יהיה עב׳[210]
ושלשה דברים וקו אב עשרה ולכן יהיה קו אע שבעה וחומש ושלש
עשיריות דבר אבל הוא כמו קו גה וקו הד ממנו שלשה וישאר קו גד
ארבעה וחומש ושלש עשיריות דבר ונכה[211] אותו על קו גב והוא דבר
ויהיה ארבעה דברים וחומש דבר ושלש עשיריות מאלגוש[212] ישוו שלשים וש
שה אדרהמיש ותשלים שלש עשיריות מאלגוש והשלמתו הוא עד שיהיה
האלגוש שלם והוא שתכהו על שלשה ושליש ותכה כל דבר שתחזיק[213] על
שלשה ושליש ויהיה אלגוש וארבע עשר דברים ישוו מאה ועשרים
אדרהמיש. ועשה כמו שאמרתי לך ויעלה הדבר ששה והוא מספר הבגדים
מאחד מהם והוא קו גב והחלק האחר הוא מה שנשאר מעשרה בגדים
והוא ארבעה והוא קו אג. ומשל.[214]

---

[207] Levey's edition (n. 56), pp. 118–121. We have changed the text in a few places and removed former readers' marginal additions since incorporated into the text.

[208] Instead of והבגדים שלקח אחד מהם the text has: בעלי הגדולה.

[209] Designation of the unknown (our x; translation of the Arabic *shay'*).

[210] The Greek number system is the origin of similar systems in Arabic (n. 48) and in Hebrew.

[211] הכה = "to beat" but—like the Arabic *ḍaraba*—in the sense of "to multiply."

[212] Designation of the unknown square, our x^2; Spanish *algo* = *cosa* (derived from *aliquod*).

[213] דבר here in the usual sense. See n. 187.

[214] Concludes a geometrical demonstration or computation (וזה מה שרצינו לבאר, Arabic *wa-ḏālika mā aradnā an nubayyina*).

ולזאת השאלה פנים אחרים והוא כמו שהוספנו[215] שנשים הבגדים שלקח אחד מהם דבר והאחרים עשרה פחות דבר. וכבר ידעת שאם נכה דבר על שומת כל בגד מהחלק האחד יהיה לו׳ אדרהמיש ואם נכה עשרה פחות דבר על שומת כל בגד מהחלק האחר יהיה כמו כן לו׳ אדרהמיש ואם תכה כל הבגדים על שומת כל בגד מהבגדים מהדבר ושלשה יהיה שבעים ושנים אדרהמיש ושלשה דברים וכאשר תחלקם על העשרה יהיה שומת כל בגד מהעשרה פחות דבר שבעה וחומש ושלש עשיריות דבר ויהיה שומת כל בגד מהבגדים מהדבר ארבעה וחומש ושלש עשיריות מדבר. ותעשה כמו שאמרתי לך במה שעבר ויעלה הדבר ששה בגדים והוא מה שלקח האחד והאחר לקח מה שנשאר מהעשרה והוא ארבעה בגדים.

וזאת השאלה היא כמו השאלה שאומר בה עשרה חלקנום לשני חלקים והכינו החלק האחד על דבר ויהיו שלשים ושש והכינו החלק האחר על דבר ושלשה ויהיו שלשים ושש. ותעשה כמו שאמרתי לך לפני זה.

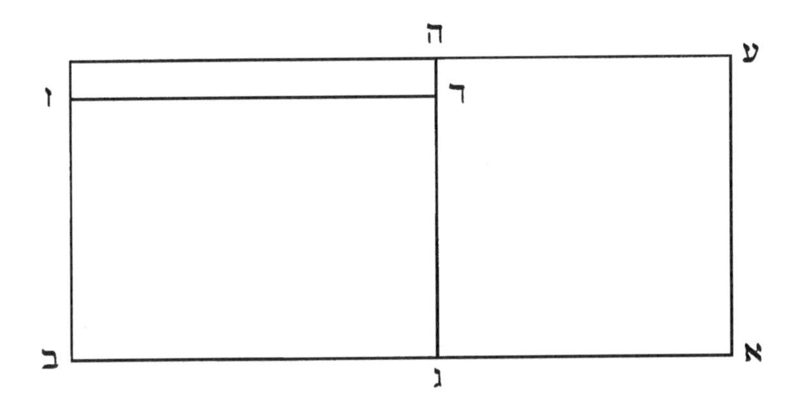

[215] The Arabic text is clearer: *wa-li-hāḏihi al-mas'alati wa-mā ashbahahā (= ašbahahā) wajh ākhar (= āḫar) ka-naḥwi mā fassartu laka.* The Latin is closer to the Arabic: *Huic autem questioni et huic consimili* (better: *consimilibus*) *est via* (omits: *altera*) *sicut exponam* (the translator thus replacing his initial, more appropriate, *exposui*) *tibi.*

Appendix E

French, German, Italian, and Provençal Texts

(1)[216] Frage: Man suche solche Zahlen x, daß ihr Quadrat doppelt genommen um 1 größer werde als ein anderes Quadrat? oder wenn man davon 1 subtrahirt ein Quadrat übrig bleibe? wie bey der Zahl 5 geschieht, deren Quadrat 25 doppelt genommen ist 50, wovon 1 subtrahirt das Quadrat 49 übrig bleibt.

Also muß $2\mathrm{xx}-1$ ein Quadrat seyn,[217] wo nach unserer Formel[218] $\mathrm{a} = -1$, $\mathrm{b} = 0$, und $\mathrm{c} = 2$, und also weder a noch c ein Quadrat ist,[219] auch läßt sich dieselbe nicht in zwey Factores auflösen, weil $\mathrm{bb} - 4\mathrm{ac} = 8$ kein Quadrat ist,[220] und daher keiner von den drey ersten Fällen statt findet.

Nach dem vierten Fall aber kann diese Formel also vorgestellt werden $\mathrm{xx}+(\mathrm{xx}-1) = \mathrm{xx}+(\mathrm{x}-1)(\mathrm{x}+1)$.[221] Hiervon werde nun die Wurzel gesetzt $\mathrm{x}+\frac{m\,(x+1)}{n}$, daher wird $\mathrm{xx}+(\mathrm{x}+1).(\mathrm{x}-1) = \mathrm{xx}+\frac{2mx\,(x+1)}{n}+\frac{mm\,(x+1)^2}{nn}$, wo sich die xx aufheben, und die übrigen Glieder durch $\mathrm{x}+1$ theilen lassen, da denn kommt $\mathrm{nnx} - \mathrm{nn} = 2\mathrm{mnx} + \mathrm{mmx} + \mathrm{mm}$ und $\mathrm{x} = \frac{mm+nn}{nn-2mn-mm}$; und weil in unserer Formel $2\mathrm{xx}-1$ nur das Quadrat xx vorkommt, so ist es gleich viel, ob die Werthe von x positiv oder negativ heraus kommen. Man kann auch sogleich $-\mathrm{m}$ anstatt $+\mathrm{m}$ schreiben, damit man bekomme $\mathrm{x} = \frac{mm+nn}{nn+2mn-mm}$. Nimmt man hier $\mathrm{m} = 1$ und $\mathrm{n} = 1$, so hat man $\mathrm{x} = 1$ und $2\mathrm{xx}-1 = 1$. Es sey ferner $\mathrm{m} = 1$ und $\mathrm{n} = 2$, so wird $\mathrm{x} = \frac{5}{7}$ und

[216] *Vollständige Anleitung zur Algebra*, St. Petersburg 1770 (1771 Lund edition used here).

[217] As with the other symbolic letters, it was usual in Euler's time to represent x^2 as xx (or, rather, xx), whereas the higher powers were already designated in the modern way.

[218] General form $\mathrm{a}+\mathrm{b\,x}+\mathrm{c\,xx}$, considered by Euler in the preceding section.

[219] Euler's cases 1 and 2, already known to Diophantus (page 35).

[220] Euler's case 3, and our case (1).

[221] Euler's theory of the *vierter Fall*, our (2), precedes this example.

$2xx - 1 = \frac{1}{49}$. Setzt man aber m = 1 und n = −2, so wird x = −5, oder x = +5 und 2xx − 1 = 49.

(2)[222] Pausa un nombre qual que tu vulhas[223] sobre lo qual tu atrobaras nombres dels quals, levadas que ne fossan las partidas prepausadas, aquell nombre resta entierament;[224] los quals nombres sobre ell trobas totz ensemps sens ell ajustatz se partiscan per 1 mentz que non son aquels en los quals se deu fer la raso;[225] et de so que vendra del partiment sustray los nombres atrobatz, car so que restara sera lo nombre d'aquell que demanda tal partida.[226] Et quant tu sustrayras lo nombre de ta positio, so que restara sera lo pretz que tu voles saber.[227]

Exemple. 3 homes volen comprar un roci, lo qual costa tant. Et cada un d'els porta tant pauc d'argent que degun d'els per si solet no lo pot comprar. Mas lo premier ditz als autres 2: Prestatz-me la $\frac{1}{2}$ de tot vostre argent, et ab lo mieu yeu comprarey lo roci. Et lo segont ditz als autres 2: Prestas-me lo $\frac{1}{3}$ de vostre argent, et yeu ab lo mieu comprarey lo roci. Lo tertz dis als autres 2: Mas me prestatz vosautres lo $\frac{1}{4}$ de vostre argent, et yeu ab lo mieu lo comprarey. Demandi que costa lo roci et que ha cascun d'els d'argent.

Resposta. Yeu pausi per mon plaser 12. Sobre lo qual nombre per aquell que demanda la $\frac{1}{2}$ atrobi 24; car qui ne levaria la $\frac{1}{2}$ restarian 12. Per lo segont, que demanda lo $\frac{1}{3}$, preni 18; car qui ne leva lo $\frac{1}{3}$ restan 12. Et per lo tertz atrobi 16; car qui ne leva lo $\frac{1}{4}$ restan 12. Ajusta aquellas 3 summas, so es 24, 18, et 16, montan 58; la qual summa devi partir per 2, que es mens 1 que non son los 3 homes, que ne ve 29. Aras per saber quant costa lo roci, leva 12 de 29, restan 17, et 17 costa lo roci. Item per saber que porta aquell que demanda la $\frac{1}{2}$, leva 24 de 29, restan 5, que porta aquell de la $\frac{1}{2}$. Item per lo segont leva 18 de 29, restan 11, que porta aquell que demanda lo $\frac{1}{3}$. Item per aquell que demanda la $\frac{1}{4}$ part, leva 16 de 29, restan 13, que porta aquell.[228]

Item 4 homes compran un rocin como desus, mas que lo quart demanda lo $\frac{1}{5}$ del argent dels autres. Demandi com desus que costa lo roci et que porta cada un d'els.

Resposta. Per mon plaser pausi 60 escutz. Per lo premier atrobi 120, per lo segont 90, per lo tertz 80, et per lo quart 75. Ajusta totas aquellas summas, so es 120, 90, 80, et 75, que montan 365; partis per 3, que es mens

[222] From the edition cited in n. 99.

[223] Choose a number, which is (in our symbols) $S - y$ (called *nombre de ta positio*).

[224] Find s_i such that $(1 - m_i)s_i = S - y$. The s_i are the numbers *trobatz* (variant writings: *trobas*, *atrobatz*).

[225] Compute $\frac{1}{n-1} \sum s_i$ $(= S)$. The *raso* is the computation.

[226] $\frac{1}{n-1} \sum s_i - s_j = x_j$, where x_j is thus the "quantity of the one requesting the part in question" (the part being m_j).

[227] $\frac{1}{n-1} \sum s_i - (S - y) = y$. This is the formula applied in all three problems.

[228] $S - y = 12$, $s_i = \{24, 18, 16\}$, $\frac{1}{n-1} \sum s_i = 29$, $y = 17$, $x_i = \{5, 11, 13\}$.

1 de 4 homes, que ne ve 121 et $\frac{2}{3}$. Aras per saber que val lo roci, leva 60 de 121 et $\frac{2}{3}$, resta 61 et $\frac{2}{3}$, et aitant val. Item per lo premier leva 120 de 121 et $\frac{2}{3}$, resta 1 et $\frac{2}{3}$, et atant portava. Item per lo segont leva 90 de 121 et $\frac{2}{3}$, restan 31 et $\frac{2}{3}$, et aitant portava. Item per lo tertz leva 80 de 121 et $\frac{2}{3}$, resta 41 et $\frac{2}{3}$; et tant porta. Item per lo quart leva 75 de 121 et $\frac{2}{3}$, restan 46 et $\frac{2}{3}$; et atant porta.[229]

Exemple de drap. Item son 5 homes que volen comprar una pessa de drap en tal maniera que lo premier demanda a totz los autres la $\frac{1}{2}$ de tot l'aur et l'argent que portan; lo segont demanda lo $\frac{1}{3}$, lo tertz demanda lo $\frac{1}{4}$, lo quart la $\frac{1}{5}$ et lo quint la $\frac{1}{6}$ part. Demandi que costa la pessa et que porta cascun d'els.

Resposta. Per mon plaser pausi 60. Aras per lo premier, yeu atrobi 120; per lo segont, 90; per lo tertz, 80; per lo quart, 75; et per lo quint, 72. Ajusta totz aquestz nombres, ses la positio que es 60, que montan 437, que deves partir per 1 mentz de los 5 homes que fan la compra; partis doncas 437 per 4, que ne ve 109 et $\frac{1}{4}$. Del qual nombre leva 60 per saber que costa la pessa, restan 49 et $\frac{1}{4}$; et aytant costa. Aras per lo premier leva-ne 120, restan 10 et $\frac{3}{4}$ mens de non res.[230] Item per lo segont leva-ne 90, resta 19 et $\frac{1}{4}$. Item per lo tertz leva-ne 80, restan 29 et $\frac{1}{4}$. Item per lo quart leva-ne 75, restan 34 et $\frac{1}{4}$. Item per lo quint leva-ne 72, restan 37 et $\frac{1}{4}$. Et aytant portava cascun d'els como restava en la sustractio de sa summa.[231]

Aras per veser la raso. Per lo premier, pren la $\frac{1}{2}$ de las summas dels autres, monta 60; del qual deves levar 10 et $\frac{3}{4}$ que lo premier ha mens de non res,[232] restan 49 et $\frac{1}{4}$, que es lo pretz de la pessa. Item per lo segont pren lo $\frac{1}{3}$ de totas las autras summas, que montan ab la sua 52 et $\frac{10}{12}$; leva-ne lo $\frac{1}{3}$ de 10 et $\frac{3}{4}$ que lo premier ha mentz de non res, restan 49 et $\frac{1}{4}$, que es lo pretz de la pessa. Item per lo tertz pren lo $\frac{1}{4}$ de totas las autras summas, montan ab la sua 51 et $\frac{15}{16}$; leva-ne lo $\frac{1}{4}$ de 10 et $\frac{3}{4}$ que ha lo premier mens de non res, restan 49 et $\frac{1}{4}$, que es lo pretz de la pessa. Item per lo quart pren lo $\frac{1}{5}$ de las autras summas, montan ab la sua 51 et $\frac{8}{20}$; leva-ne lo $\frac{1}{5}$ de 10 et $\frac{3}{4}$ que ha lo premier mens de non res, resta 49 et $\frac{1}{4}$, que es la valor de la pessa. Item per lo derrier pren la $\frac{1}{6}$ part de las autras summas, montan ab la sua 51 et $\frac{1}{24}$; leva-ne la $\frac{1}{6}$ part de 10 et $\frac{3}{4}$ que ha lo premier mens de non res, restan 49 et $\frac{1}{4}$, que es la valor de la pessa del drap.

---

[229] $S - y = 60$, $s_i = \{120, 90, 80, 75\}$, $\frac{1}{n-1}\sum s_i = 121 + \frac{2}{3}$, $y = 61 + \frac{2}{3}$, $x_i = \{1 + \frac{2}{3}, 31 + \frac{2}{3}, 41 + \frac{2}{3}, 46 + \frac{2}{3}\}$.

[230] First negative solution accepted ("10 and $\frac{3}{4}$ less than no thing remain").

[231] $S - y = 60$, $s_i = \{120, 90, 80, 75, 72\}$, $\frac{1}{n-1}\sum s_i = 109 + \frac{1}{4}$, $y = 49 + \frac{1}{4}$, $x_i = \{-(10 + \frac{3}{4}), 19 + \frac{1}{4}, 29 + \frac{1}{4}, 34 + \frac{1}{4}, 37 + \frac{1}{4}\}$.

[232] "10 and $\frac{3}{4}$ that the first has less than no thing."

(3)[233] L'on doit savoir que qui adjouste ou soustrait 0 avec aulcun nombre, l'addition ou soustraction ne augmente ne diminue; et qui adjouste ung moins avec ung aultre nombre ou qui d'icellui le soustrayt, l'addition se diminue et la soustraction croist.[234] Ainsi comme qui adjouste moins 4 avec 10, l'addition monte 6; et qui de 10 en soustrait moins 4, il reste 14. Et quant l'on dit moins 4, c'est comme si une personne n'avoit riens et qu'il deust encores 4. Et quant on dit 0, c'est rien simplement. (...)

Je veulx trouver cinq nombres de telle condition que tous ensemble sans le premier montent 120, et sans le second montent 90, sans le tiers facent 80, sans le quart 75, et sans le quint 72.[235]

Pour iceulx trouver je assemble 120, 90, 80, 75 et 72, font 437, que je partiz par 4, qui sont 1 moins de cinq nombres, et m'en vient 109 $\frac{1}{4}$. Desquelz je lyeve 120, 90, 80, 75 et 72. Et me reste moins 10 $\frac{3}{4}$ pour le premier, 19 $\frac{1}{4}$ pour le second, 29 $\frac{1}{4}$ pour le tiers, 34 $\frac{1}{4}$ pour le quart, et 37 $\frac{1}{4}$ pour le quint. Ce sont les nombres que je demandoye.

Encores, je veulx trouver cinq nombres de telle nature que tous ensemble sans le premier facent 120, sans le second 180, sans le tiers 240, sans le quart 300, et sans le quint 360.

Et pour iceulx trouver je assemble tous ces cinq nombres, et montent 1200; que je divise par 4, et m'en vient 300; desquelz je soustraiz les cinq nombres dessusdicts, c'est assavoir 120, 180, 240, 300, et 360, et me restent 180, 120, 60, 0, et moins 60, qui sont les cinq nombres que je desiroye.

(4)[236] Quando che'l cubo con le cose appresso
Se agguaglia a qualche numero discreto
Trovan dui altri differenti in esso.
Dapoi terrai questo per consueto
Che'l lor produtto sempre sia eguale
Al terzo cubo delle cose neto.
El residuo poi suo generale
Delli lor lati cubi ben sottratti
Varrà la tua cosa principale.

In el secondo de cotesti atti
Quando che'l cubo restasse lui solo
Tu osservarai quest'altri contratti.
Del numer farai due tal part'a volo
Che l'una in l'altra si produca schietto
El terzo cubo delle cose in stolo.
Delle qual poi, per commun precetto,

---

[233] Marre's edition (n. 100), pp. 641–42; or MS Bibliothèque Nationale de France, fr. 1346, fol. 35^v–36^r.

[234] If $a,\ b > 0$, so $a \pm 0 = a,\ \ a + (-b) < a,\ \ a - (-b) > a.$

[235] $S - x_j = \frac{1}{n-1} \sum s_i - x_j = s_j = \{120, 90, 80, 75, 72\}$.

[236] *Quesiti et inventioni diverse* (n. 112), fol. 120^v.

Torrai li lati cubi insieme gionti
Et cotal somma sarà il tuo concetto.

El terzo poi de questi nostri conti
Se solve col secondo se ben guardi
Che per natura son quasi congionti.

Questi trovai, & non con passi tardi
Nel mille cinquecente, quatro e trenta
Con fondamenti ben sald' e gagliardi
Nella città dal mar' intorno centa.

(5)[237] Dimostratione di Cubo eguale a Tanti e numero in superficie piana.[238]

Sia $1\overset{3}{\smile}$ eguale a $6\overset{1}{\smile}$ p. 4 e sia la q la unità. Tirisi la me e faccisi ml che sia pari alla q, cioè sia 1, et lf 6, cioè quanto è il numero delli Tanti. E sopra detta lf si faccia un paralellogramo[239] che sia 4 di superficie, cioè quanto il numero; e sarà il paralellogramo abf. Poi allonghisi la ab sino in d ed al sino in r. Poi habbiansi due squadri, delli quali l'uno si ponga con l'angolo sopra la linea r[240] e che l'uno delle braccia tocchi la estremità m, il qual squadro si alzi o abbassi tanto che tirato dall'angolo del squadro una linea che tocchi la estremità f che vada a toccare la bd in tal luogo che, mettendo un altro squadro con l'angolo al detto toccamento et con l'uno delle braccia sopra la da vadi a intersegare il braccio dell'altro squadro nella linea f[241]. E fatto questo, dico che la linea ch'è dal punto l sino all'angolo del squadro è la valuta del Tanto.

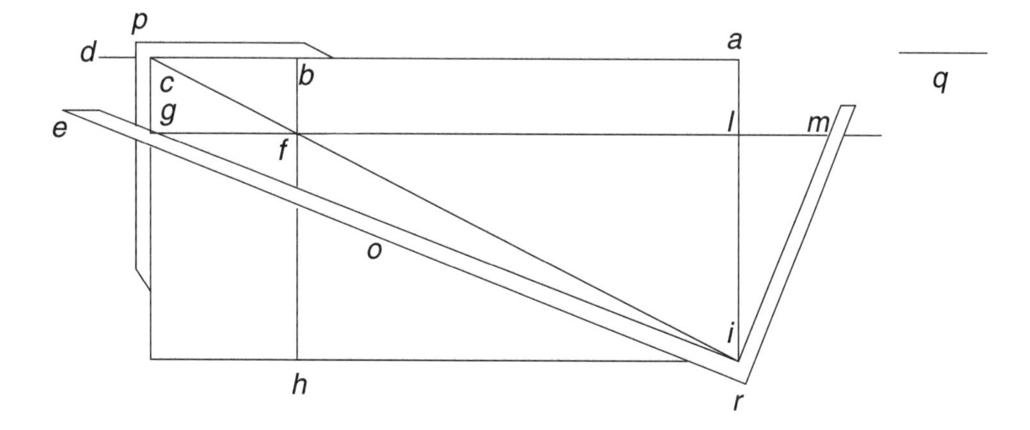

[237] Bortolotti's edition (n. 109), pp. 228–29 (with minor changes).

[238] Normally the solution of a cubic equation is illustrated by means of a stereometric figure.

[239] *paralellogramo* (= Bombelli's writing) is also used for rectangles.

[240] Bombelli sometimes designates a line by a single point on it.

[241] That is, the extension of lf. This other *squadro* in p thus serves to bring about the vertical alignment of c (*toccamento*) and g.

E lo provo in questo modo. Presuposto che si habbia alzato e abbassato lo squadro talmente che in i tirando la if sino in c e che il braccio dello squadro p tagliassi con l'altro squadro in g suso la linea ge, fatto questo, dico la linea li essere la valuta del Tanto. Perchè, essendo la li $1\overset{1}{\smile}$ et ml 1, la lg sarà $1\overset{2}{\smile}$, perchè tanto può la ml in lg quanto li in se stessa —per esser l'angolo i retto; il paralellogramo ilg sarà un cubo[242]. Ed il paralellogramo ilf sarà $6\overset{1}{\smile}$ perchè il è $1\overset{1}{\smile}$ et lf 6; et il paralellogramo hfg sarà 4, perch'è pari al paralellogramo alf ch'era 4. Et essendo ilg tutto insieme $6\overset{1}{\smile}$ e 4, e per l'altra ragione è provato essere $1\overset{3}{\smile}$, dunque $1\overset{3}{\smile}$ sarà eguale a $6\overset{1}{\smile}$ p. 4, e la il sarà $1\overset{1}{\smile}$.

Che per la agguagliatione insegnata[243] la li sarà R.q. 3 p. 1, la lg sarà 4 p. R.q. 12, [la fg sarà R.q. 12 m. 2,][244] il paralellogramo ilg sarà R.q. 108 p. 10. Et il paralellogramo ilf sarà R.q. 108 p. 6 per essere la linea il R.q. 3 p. 1 e la lf 6; il paralellogramo hfg è 4, che gionto insieme con R.q. 108 p. 6 fa R.q. 108 p. 10, ch'è pari al cubo ilg —come fu proposto.

---

[242] Rather: sarà $1\overset{3}{\smile}$.

[243] Now verifying the equality with the numerical values: with $il = \sqrt{3} + 1$, we have first $lg = 4 + \sqrt{12}$, so $ilg = \sqrt{108} + 10$; second, since $ilf = il \cdot lf = \sqrt{108} + 6$ and $hfg = 4$ (by construction), their sum is $\sqrt{108} + 10$.

[244] This remark is not necessary.

Index

'Abd al-Ḥamīd ibn Turk 59–61
Abel, N. 140–41
Abū Kāmil 46–49, 55, 61, **63–79**, 94, 96, 112
Abū'l-Jūd 85–86
Akkadian 1
Alcuin 78, 93, 104
algebra (origin of the name) 34
algebraic identities *see* identities
algebraic language and symbols *see* symbolism
algo 163 (*n.* 212)
algorithmus 56 (origin of the name), 138*n*
Almoravides 100
Amthor, A. 31
Anbouba, A. 46*n*
Apollonius 54
application of areas 79
Archimedes 17, 27–28, 31, 54, 82–83, 85. *See also* cattle problem
Aristotle 71*n*
Arrighi, G. 120*n*

Bachet de Méziriac, C. G. 38*n*, 49
Bahā' al-Dīn al-'Āmilī 63
bayt al-ḥikma 53
bird problems (*masā'il al-ṭuyūr*) 75–79
Bīrūnī 53, 86
Blume, F. 21*n*
Boethius 93
Bolognetti, P. 129–30
Bombelli, R. 49, 123–24, **135–39**, 169*n*
Boncompagni, B. 95*n*
Bortolotti, E. 124*n*, 130*n*, 132*n*, 138*n*, 169*n*
Bruins, E. 148*n*
Bubnov, N. 21*n*, 148*n*
Busard, H. 94*n*
Byzantine contributions 18, 32, 94–95, 103, 106*n*, 110, 112

Cajori, F. 125*n*
Cantor, M. v.
Cardano, G. 123, 126–27, **131–35**, 136
casus irreducibilis 139
cattle problem of Archimedes 27–31
censo 123
census 96, 123, 157 (*n.* 170)
Chinese contributions 78, 104, 112
Christianidis, J. 5*n*, 44*n*
Chuquet, N. 116–17, 123, 125
Colla, G. 132, 134
Compendion del art de algorisme 116–17
complex numbers 63, 89, **135–40**
conic sections 81, 82, 85, 88–89, 160*n*
continued fractions 31*n*
cosa 118, 123–24, 130
cubic equation 63, 80–83, 84–86, **88–91**, 95, 118–21, **125–28**, **129–34**, 135–40, 140*n*, 168–70
cubo 123
cubus 96, 123

Dardi of Pisa 119–20
decagon 84
decimal number system 1–4
Delian problem *see* duplication of the cube
Descartes, R. 125, 139
difference of two squares 36, 39–42

Diophantus and his *Arithmetica* 17, 24–26, **31–46**, 49–50, 54–55, 110, 112, 115–16, 158 (*n.* 181), 165*n*
Djebbar, A. 96*n*
duplication of the cube 81–82, 85

van Egmont, W. 118*n*, 119*n*
enneagon 83, 86
equations (simplification of) 9, 34, 62; (six standard forms) 57, 60–61, 64, 69–70, 88, 88*n*; (systems of) 1, 4–9, 11–15, 25–31, 40–49, 61–63, 67–70, 75–79, 98, 100–1, 103–19, 126, 130, 134–36. *See also* quadratic, cubic, quartic, quintic; geometric solving; indeterminate; inconsistent, unsolvable
Eratosthenes 27
Euclid and his *Elements* 54–55, 64–66, 71, 79–81, 83, 93, 96, 98–99, 152
Euler, L. 24, 49–51, 124, 165*n*

de Fermat, P. 31, 38, 44, 49, 83*n*
Ferrari, L. 132*n*, 133–34
dal Ferro, S. 129–31, 134
Fibonacci, L. 63, 95, **101–15**, 117
Fibonacci sequence 102–3
Fior, A. M. 130–131, 133, 134*n*
Flügel, G. 61*n*
Folkerts, M. 78*n*
della Francesca, P. 120
Franci, R. 120*n*
fundamental theorem of algebra 139–140

Galois, E. 141
Gauss, C. F. 49, 83*n*, 139
geometric constructions by compass and straightedge 72–73, 81–84
geometric illustrations of solutions 54, 57–60, 65–69, 83–87, 96, 98–101, 136–38, 163 (*n.* 214), 169 (*n.* 238)
geometric solving of equations 73, 79–91, 136–38
Gerardi, P. 118–19
Gericke, H. 78*n*
Girard, A. 139
golden ratio 84
Greek Anthology 25, 104
Greek number systems 4, 17–18, 55, 163 (*n.* 210)
Greek science, transmission of 3–4, 44, 46, 49, 53–55, 78, 93–94

Ḥājjī Khalīfa 61*n*
Halma, N. 4*n*
Heath, T. v, 37*n*
Heiberg, J. 4*n*, 22*n*, 85*n*
Henry, C. 38*n*
heptagon 83, 85
Heron of Alexandria 22, 54
hexagon 83
Hippocrates of Chios 81–82
Hogendijk, J. 85*n*
Høyrup, J. 5*n*
Hypatia 32
hyperbola 88–91

Iamblichus 26–27
Ibn al-Haytham 83
Ibn Khaldūn 63
identities, use of (in algebra) 1, **11–12**, 13, 15–16, **19**, 20–25, 65–66, 98–99, **126**, **130**
al-ikmāl 62, 69
imaginary numbers 135–140. *See also* complex numbers
impossible, equation or problem *see* unsolvable
inconsistent equations 108–9
indeterminate equations and systems 22–25, 27–31, 33, 35–51, 64, 75–79, 109, 111, 114, 116
Indian contributions 53–56, 78, 104, 110; (number system) 54, 95, 101
infinite number of solutions *see* solutions
interpolations or lacunas in manuscripts 23*n*, 43*n*, 44, 45 (*n.* 38), 147 (*n.* 125), 148 (*n.* 129), 150 (*n.* 134, 135), 153 (*n.* 151), 154 (*n.* 156), 158 (*n.* 174), 163 (*n.* 207)
irrational numbers 63, 64, **69–72**, 73–75, 96, 135.
Islamic science, transmission of 56, 63, 65*n*, 93–97, 104, 110, 112

al-jabr 34, 46, 62, 63*n*, 159*n*
John of Seville 96

ka‘b 55. *See also: cubus*
Karajī 46
Karpinski, L. 26*n*
Kennedy, E. 53*n*
Khayyām, ‘U. 81, **88–91**
al-Khāzin 83

al-Khwārizmī 55, **56–63**, 64, 79, 88*n*, 93–94
Kummer, E. 38

Lachmann, K. 21*n*
Leibniz, W. 124
Lessing, G. E. 31*n*
Levey, M. 64*n*, 163*n*
Liber abaci 101–15
Liber mahameleth 93, 95–102
Libri, G. 57*n*, 94*n*
Lorch, R. 64*n*

māl 55. *See also: census*
Manitius, K. 4*n*
Marre, A. 116*n*, 168*n*
Mellema, E. E. L. 49
Menaechmus 82
Menelaus 54
Menninger, K. 18*n*
Mesopotamian number system 1–4
Mesopotamian science, transmission of 3–4, 17, 19, 25, 53
Migne, J. 78*n*
al-muqābala 34, 46, 63*n*, 159*n*
Muscus 103

negative numbers 24–25, 51, **103**, **115–18**, 135, 139, 167*n*
Nesselmann, G. 63*n*
Newton, I. 88
Nicomachus 54, 93
nonagon *see* enneagon
numbers *see* irrational, negative, complex; triangular; sum of squares, difference of two squares; decimal, sexagesimal
number systems and symbols *see* Greek, Indian, Mesopotamian
number theory 32, 36. *See also* difference of two squares, sum of squares

Omar Khayyam *see* Khayyām.
Oughtred, W. 124

Pacioli, L. 95, 117–19, 123, 135
parabola 88–90
Pazzi, A. 49
Pell's equation 31*n*
pentagon 73–75, 84
polygons, constructible 73, 83–87
polygons *see individual names*
powers of the unknown, designation of 32–33, 55–56, 96, 123, 163 (*n.* 212)
problems *see* bird problems, cattle problem, ten problems; indeterminate, recreational, unsolvable
Ptolemy 3–4, 54, 93
Pythagorean(s) 26, 30, 71*n*
Pythagorean theorem 19

quadratic equation 1, **9–11**, 33, 35, 57–60, 65–67, 68–71, 73–75, **79–81**, 83–84, 88, 99, 118–19, 129, 135–36, 140*n*, 158 (*n.* 175)
quartic equation 95, 120, **128–29**, **134–35**, 140, 140*n*, 158 (*n.* 179)
quintic equation 87, 140–41
Qusṭā ibn Lūqā 37*n*

al-radd 62
radicals, solution by 140–41
Rahn, J. 124
Recorde, R. 125
recreational problems 25, 64, 93, 97, 102
Regiomontanus, J. 49
Reich, K. 78*n*
res 96, 123
Robbins, F. 26*n*
Roman surveyors 21
Rosen, F. 57*n*, 61*n*, 157*n*
Rudorff, A. 21*n*
Ruffini, P. 140

Sayılı, A. 59*n*
Schmidt, G. 22*n*
sexagesimal number system 1–4
Sezgin, F. 32
shay' 55, 158*n*, 159*n*, 163*n*. *See also: res*
Sigler, L. 95*n*
al-Sijzī 85
de Slane, M. 63*n*
solutions, numerical (positive) 9–10, 35–36, 50–51, 59–60, 73, 80, 88–90, 115–16, 129, 136; (rational) 9, 23–24, 33–38, 47–48, 63; (negative) 24, 51, 103, 108, 114–18, 129, 139, 167*n*; (irrational) 63, 70–72 (*see also* irrational numbers); (equal to zero) 76, 79, 107–8, 117; none 75 (*see also* inconsistent, unsolvable); (complex) 63, 89, 127, 129, 136, 139; (infinite number of) 35–36; (integral) 38, 63 (*see also* bird

problems, cattle problem); (enumeration of) 75–79. *See also* radicals
squaring the circle 119
Stevin, S. 49
sum of squares 36–39, 41–47
Sumerians 1, 4
surveyors 21
Suter, H. 75*n*
Symbolism and algebraic language 5 (*n.* 12), 14*n*, 20*n*, 21 (*n.* 25), 25–26, **32–33**, 45 (*n.* 39), **55–56**, 62, 71–72, 96, 103, 105, **123–25**, 130, 138–39, 158 (*n.* 172, 175, 180, 181), 163 (*n.* 209, 212), 165 (*n.* 217)
systems of equations *see* equations

tables, Mesopotamian numerical 3, 6*n*
Tannery, P. 37*n*, 38*n*, 149*n*
tanto 124, 169–170
Tartaglia, N. **130–134**
Taylor, R. 38
ten problems (*masā'il al-'asharāt*) 61, **62–63**, 67–70, 117–19, 135
Theon of Alexandria 31–32
thing (designation of the unknown) *see*: *cosa*, *res*, *shay'*, *tanto*
Thureau-Dangin, F. 5*n*, 143*n*
Thymaridas, rule of 26–27, 103, 117
Toomer, G. 4*n*, 44*n*, 55*n*
translation of mathematical works 32, **54–55**, 56, 57*n*, 64–65*n*, 71–72, **93–94**
transmission of mathematical texts *see* Greek, Islamic, Mesopotamian; *see also* translation, interpolations
triangles, right-angled in rational numbers 19–25, 38
triangular number 30
trigonometry 53–54, 139
trisection of an angle 85
Tropfke, J. v, 78*n*, 103*n*, 104*n*, 110*n*, 112*n*, 125*n*
von Tschirnhaus, E. 140

'Umar Khayyām *see* Khayyām
unknown (auxiliary) 4, 7-8, 13, 26, 33, 45; (designation of) 25–26, 45 (*n.* 39), 55, 56, 76, 96, 105, 158 (*n.* 181), 163 (*n.* 209). *See also* powers
unsolvable, considered as 36, 47, 63, 75, 90–91, 106–9, 111–12, 114–15, 135–36

Viète, F. (and Vieta's formulas) 80
Vogel, K. v, 5*n*, 78*n*, 103*n*

Wiles, A. 38
Winter, J. 26*n*
Woepcke, F. 46*n*, 83*n*, 86*n*, 160*n*

Xylander, G. 49

zero 2, 5, 18, 101*n*

Titles in This Series

27 **Jacques Sesiano,** An Introduction to the History of Algebra: Solving Equations from Mesopotamian Times to the Renaissance, 2009

26 **A. V. Akopyan and A. A. Zaslavsky,** Geometry of conics, 2008

25 **Anne L. Young,** Mathematical Ciphers: From Caesar to RSA, 2006

24 **Burkard Polster,** The shoelace book: A mathematical guide to the best (and worst) ways to lace your shoes, 2006

23 **Koji Shiga and Toshikazu Sunada,** A mathematical gift, III: The interplay between topology, functions, geometry, and algebra, 2005

22 **Jonathan K. Hodge and Richard E. Klima,** The mathematics of voting and elections: A hands-on approach, 2005

21 **Gilles Godefroy,** The adventure of numbers, 2004

20 **Kenji Ueno, Koji Shiga, and Shigeyuki Morita,** A mathematical gift, II: The interplay between topology, functions, geometry, and algebra, 2004

19 **Kenji Ueno, Koji Shiga, and Shigeyuki Morita,** A mathematical gift, I: The interplay between topology, functions, geometry, and algebra, 2003

18 **Timothy G. Feeman,** Portraits of the Earth: A mathematician looks at maps, 2001

17 **Serge Tabachnikov, Editor,** Kvant Selecta: Combinatorics, I, 2001

16 **V. V. Prasolov,** Essays on number and figures, 2000

15 **Serge Tabachnikov, Editor,** Kvant Selecta: Algebra and analysis. II, 1999

14 **Serge Tabachnikov, Editor,** Kvant Selecta: Algebra and analysis. I, 1999

13 **Saul Stahl,** A gentle introduction to game theory, 1999

12 **V. S. Varadarajan,** Algebra in ancient and modern times, 1998

11 **Kunihiko Kodaira, Editor,** Basic analysis: Japanese grade 11, 1996

10 **Kunihiko Kodaira, Editor,** Algebra and geometry: Japanese grade 11, 1996

9 **Kunihiko Kodaira, Editor,** Mathematics 2: Japanese grade 11, 1997

8 **Kunihiko Kodaira, Editor,** Mathematics 1: Japanese grade 10, 1996

7 **Dmitry Fomin, Sergey Genkin, and Ilia Itenberg,** Mathematical circles (Russian experience), 1996

6 **David W. Farmer and Theodore B. Stanford,** Knots and surfaces: A guide to discovering mathematics, 1996

5 **David W. Farmer,** Groups and symmetry: A guide to discovering mathematics, 1996

4 **V. V. Prasolov,** Intuitive topology, 1995

3 **L. E. Sadovskiĭ and A. L. Sadovskiĭ,** Mathematics and sports, 1993

2 **Yu. A. Shashkin,** Fixed points, 1991

1 **V. M. Tikhomirov,** Stories about maxima and minima, 1990